· THE OXFORD SHERLOCK HOLMES ·

General Editor
Owen Dudley Edwards

The Case-Book of Sherlock Holmes

ARTHUR CONAN DOYLE

The Case-Book of Sherlock Holmes

Edited with an Introduction by

W. W. Robson

Oxford New York

OXFORD UNIVERSITY PRESS

1993

Oxford University Press, Walton Street, Oxford OX2 6DP
Oxford New York Toronto
Delhi Bombay Calcutta Madras Karachi
Kuala Lumpur Singapore Hong Kong Tokyo
Nairobi Dar es Salaam Cape Town
Melbourne Auckland Madrid
and associated companies in
Berlin Ibadan

Oxford is a trade mark of Oxford University Press

British Library Cataloguing in Publication Data
Data available

Library of Congress Cataloging in Publication Data
Data available

ISBN 0-19-212311-4
ISBN 0-19-212329-7 (set)

1 3 5 7 9 10 8 6 4 2

Typeset by Pure Tech Corporation, Pondicherry, India
Printed in Great Britain
on acid-free paper by
BPCC Paperbacks Ltd
Aylesbury, Bucks

CONTENTS

ACKNOWLEDGEMENTS

I WISH to thank the Librarian and staff of the National Library of Scotland, of Edinburgh University Library, and of Edinburgh Central Library (especially its Reference Division).

I must also express my gratitude to Professor Bonnie S. McDougall for her invaluable judgement and information respecting Chinese pottery as discussed in 'The Illustrious Client', to Professor David Whitteridge, FRS, FRSE, for very helpful biographical detail on the career of William Rutherford which threw some light on the antecedents of 'The Creeping Man', to Professor Rosalind Mitchison, and to Professor Christopher Ricks.

My debt to Hugh Robson for his pains in preparing this work for the press is very difficult to express with a sufficiency of thanks. My debt to Anne Robson can never receive sufficient acknowledgement.

WWR

GENERAL EDITOR'S PREFACE TO THE SERIES

ARTHUR CONAN DOYLE told his *Strand* editor, Herbert Greenhough Smith (1855–1935), that 'A story always comes to me as an organic thing and I never can recast it without the Life going out of it.'[1]

On the whole, this certainly seems to describe Conan Doyle's method with the Sherlock Holmes stories, long and short. Such manuscript evidence as survives (approximately half the stories) generally bears this out: there is remarkably little revision. Sketches or scenarios are another matter. Conan Doyle was no more bound by these at the end of his literary life than at the beginning, whence scraps of paper survive to tell us of 221B Upper Baker Street where lived Ormond Sacker and J. Sherrinford Holmes. But very little such evidence is currently available for analysis.

Conan Doyle's relationship with his most famous creation was far from the silly label 'The Man Who Hated Sherlock Holmes': equally, there was no indulgence in it. Though the somewhat too liberal Puritan Micah Clarke was perhaps dearer to him than Holmes, Micah proved unable to sustain a sequel to the eponymous novel of 1889. By contrast, 'Sherlock' (as his creator irreverently alluded to him when not creating him) proved his capacity for renewal 59 times (which Conan Doyle called 'a striking example of the patience and loyalty of the British public'). He dropped Holmes in 1893, apparently into the Reichenbach Falls, as a matter of literary integrity: he did not intend to be written off as 'the Holmes man'. But public clamour turned Holmes into an economic asset that could not be ignored. Even so, Conan Doyle could not have continued to write about

[1] Undated letter, quoted by Cameron Hollyer, 'Author to Editor', *ACD—The Journal of the Arthur Conan Doyle Society*, 3 (1992), 19–20. Conan Doyle's remark was probably *à propos* 'The Red Circle' (*His Last Bow*).

Holmes without taking some pleasure in the activity, or indeed without becoming quietly proud of him.

Such Sherlock Holmes manuscripts as survive are frequently in private keeping, and very few have remained in Britain. In this series we have made the most of two recent facsimiles, of 'The Dying Detective' and 'The Lion's Mane'. In general, manuscript evidence shows Conan Doyle consistently underpunctuating, and to show the implications of this 'The Dying Detective' (*His Last Bow*) has been printed from the manuscript. 'The Lion's Mane', however, offers the one case known to us of drastic alterations in the surviving manuscript, from which it is clear from deletions that the story was entirely altered, and Holmes's role transformed, in the process of its creation.

Given Conan Doyle's general lack of close supervision of the Holmes texts, it is not always easy to determine his final wishes. In one case, it is clear that 'His Last Bow', as a deliberate contribution to war propaganda, underwent a ruthless revision at proof stage—although (as we note for the first time) this was carried out on the magazine text and lost when published in book form. But nothing comparable exists elsewhere.

In general, American texts of the stories are closer to the magazine texts than British book texts. Textual discrepancies, in many instances, may simply result from the conflicts of sub-editors. Undoubtedly, Conan Doyle did some re-reading, especially when returning to Holmes after an absence; but on the whole he showed little interest in the constitution of his texts. In his correspondence with editors he seldom alluded to proofs, discouraged ideas for revision, and raised few—if any—objections to editorial changes. For instance, we know that the *Strand*'s preference for 'Halloa' was not Conan Doyle's original usage, and in this case we have restored the original orthography. On the other hand, we also know that the *Strand* texts consistently eliminated anything (mostly expletives) of an apparently blasphemous character, but in the absence of manuscript confirmation we have normally been unable to restore what were probably stronger original versions. (In any case, it is perfectly possible that Conan Doyle, the consummate professional, may have come to exercise self-censorship in the certain knowledge that editorial changes would be imposed.)

Throughout the series we have corrected any obvious errors, though these are comparatively few: the instances are at all times noted. (For a medical man, Conan Doyle's handwriting was commendably legible, though his 'o' could look like an 'a'.) Regarding the order of individual stories, internal evidence makes it clear that 'A Case of Identity' (*Adventures*) was written before 'The Red-Headed League' and was intended to be so printed; but the 'League' was the stronger story and the *Strand*, in its own infancy, may have wanted the series of Holmes stories established as quickly as possible (at this point the future of both the Holmes series and the magazine was uncertain). Surviving letters show that the composition of 'The Solitary Cyclist' (*Return*) preceded that of 'The Dancing Men' (with the exception of the former's first paragraph, which was rewritten later); consequently, the order of these stories has been reversed. Similarly, the stories in *His Last Bow* and *The Case-Book of Sherlock Holmes* have been rearranged in their original order of publication, which—as far as is known—reflects the order of composition. The intention has been to allow readers to follow the fictional evolution of Sherlock Holmes over the forty years of his existence.

The one exception to this principle will be found in *His Last Bow*, where the final and eponymous story was actually written and published after *The Valley of Fear*, which takes its place in the Holmes canon directly after the magazine publication of the other stories in *His Last Bow*; but the removal of the title story to the beginning of the *Case-Book* would have been too radically pedantic and would have made *His Last Bow* ludicrously short. Readers will note that we have already reduced the extent of *His Last Bow* by returning 'The Cardboard Box' to its original location in the *Memoirs of Sherlock Holmes* (after 'Silver Blaze' and before 'The Yellow Face'). The removal of 'The Cardboard Box' from the original sequence led to the inclusion of its introductory passage in 'The Resident Patient': this, too, has been returned to its original position and the proper opening of 'The Resident Patient' restored. Generally, texts have been derived from first book publication collated with magazine texts and, where possible, manuscripts; in the case of 'The Cardboard Box' and 'The Resident Patient', however, we have employed the *Strand* texts, partly because of the restoration of the latter's opening, partly to give readers a flavour of the magazine in which the Holmes stories made their first, vital conquests.

In all textual decisions the overriding desire has been to meet the author's wishes, so far as these can be legitimately ascertained from documentary evidence or application of the rule of reason.

One final plea. If you come to these stories for the first time, proceed now to the texts themselves, putting the introductions and explanatory notes temporarily aside. Our introductions are not meant to introduce: Dr Watson will perform that duty, and no one could do it better. Then, when you have mastered the stories, and they have mastered you, come back to us.

OWEN DUDLEY EDWARDS

University of Edinburgh

INTRODUCTION

Few can doubt that the eponymous Epilogue to *His Last Bow* was originally meant to be the final appearance of Sherlock Holmes. No better ending to the series could have been devised than Watson's characteristic misunderstanding of Holmes's 'There's an east wind coming, Watson', followed by Holmes's two-edged compliment to him as 'the one fixed point in a changing age'. The year 1914 saw the end of an era and the beginning of a new world in which they were both incongruous survivors. Yet there were to be more stories. Similarly *The Tempest* has often been seen as Shakespeare's farewell to the stage, but he seems to have gone on writing plays, thereby giving another example of the 'positively last appearance' common among members of his profession.

What were Conan Doyle's reasons for resuming the stories, after so emphatic a finale? No doubt there was a financial motive, and not necessarily a selfish one: he needed the money for various good causes, above all his work for Spiritualism. But the stimulus to write the *Case-Book* seems to have come from the films based on Sherlock Holmes made by the Stoll Company in 1921–3. (There is a full discussion of them in Michael Pointer's *The Public Life of Sherlock Holmes* (1975)). The first series (1921), directed by Maurice Elvey (1887–1967) and entitled *The Adventures of Sherlock Holmes*, consisted of fifteen episodes chosen at random from the short stories; Elvey followed it with a full-length *Hound of the Baskervilles*. When Elvey went to Hollywood, George Ridgwell took over the direction of the second and third series. The second series (1922) was called *The Further Adventures of Sherlock Holmes*, and again fifteen episodes were taken at random from the short stories. The third series, *The Last Adventures of Sherlock Holmes* (1923), was once again based on fifteen episodes from the short stories. Late in 1923 Elvey returned from America and produced the last of the Stoll

Sherlock Holmes films, taken from another of the long stories (*The Sign of* [*the*] *Four*).

Doyle was enthusiastic about these films. 'The impression that Holmes was a real person of flesh and blood', he wrote, 'may have been intensified by his frequent appearances on the stage.' Not all Holmes's admirers would agree with this. Some readers who are unconvinced by Holmes plays, films, and television series get an overwhelming sense of reality from Holmes on the page. (The mediation of Watson is an important element here.) But there was no question of Doyle cynically selling out to the film industry. The Stoll films were mostly close adaptations of the original stories. Doyle co-operated in the making of the series and was immensely impressed by the performance of Eille Norwood (1861–1948) as Holmes, declaring in *Memories and Adventures* (p. 126) 'He has that rare quality which can only be described as glamour, which compels you to watch an actor eagerly even when he is doing nothing.' Norwood fitted Doyle's own notion of Holmes: 'He has the brooding eye which excites expectation and he has also a quite unrivalled power of disguise.' Norwood's calm, restrained style as an actor also struck observers. While many of the other actors went in for exaggeration and grimace, he remained quiet and thoughtful, storing away observations and deductions, which he would reveal dramatically at the denouement.

One difference between the Stoll series and earlier Holmes films was the 1920s setting, which troubled Doyle: 'My only criticism of the films is that they introduce telephones, motor cars and other luxuries of which the Victorian Holmes never dreamed.' But the decision of the producer, which Doyle had queried, was to influence the later stories, in which telephones and motor cars appear, and which are, in subtle ways, more modern than the earlier stories. Doyle clearly resisted the temptation to make them period pieces.

Norwood's performance suggested the Sherlock Holmes play *The Crown Diamond*, supposedly commissioned by Oswald Stoll, and first staged in May 1921 in Bristol and then in London. The play and the films helped to stimulate a revival of interest in Holmes, and the editor of the *Strand* was eager for more stories. In September 1921 a revised version of the play appeared in the magazine as 'The Adventure of the Mazarin Stone'. Doyle complained of the lack of good ideas for stories, so the editor, Greenhough Smith, offered one, based on a German who tied a stone to a revolver, hung it over the side of a bridge, then shot himself, the weapon vanishing into the water. Doyle began writing the story as 'The Adventure of the Second Chip', but the title gave away too much of the plot, and so was changed to 'The Problem of Rushmore Bridge' and eventually to 'The Problem of Thor Bridge'. It was followed by four more at irregular intervals, and finally by a series of six.

The *Case-Book* was widely judged to be inferior to the early stories. To John Gore (1885–1983), who had criticized them adversely, Doyle wrote in November 1926: 'I wonder whether the smaller impression which they produce upon you may not be due to the fact that we become blasé and stale ourselves as we grow older . . . I test the Holmes stories on fresh young minds and I find that they stand the test well.' He thought highly of at least two of the later stories, 'The Illustrious Client' and 'The Lion's Mane', but many critics have written off the *Case-Book* as a waste-paper basket, an exhausted performance.

These later stories *are* significantly different from most of the earlier ones. What this difference is may be suggested by a passage in 'The Greek Interpreter' (*Memoirs*) quoted by the American critic Edmund Wilson (1895–1972):

I stood gazing round and wondering where on earth I might be, when I saw someone coming towards me in the darkness. As he came up to me I made out that it was a railway porter.

'Can you tell me what place this is?' I asked.

'Wandsworth Common,' said he.

'Can I get a train into town?'

'If you walk a mile or so, to Clapham Junction,' said he, 'you'll just be in time for the last to Victoria.'

Whatever outlandish adventures may befall the interpreter the trains will run on time; God's in his heaven and all's right with the Establishment. Most of the pre-*Case-Book* stories produce a final effect of cosiness, ending with a crossing of the t's and dotting of the i's in the snug rooms in Baker Street. Edmund Wilson remarks another quality that makes for cheerfulness in the stories, the note of 'irresponsible comedy', as of some father's rigmarole for children, that pervades the whole series. In some moods the characterization of Holmes is irrepressibly comic.

'I have lost my thumb, and I have lost a fifty-guinea fee, and what have I gained?'

'Experience,' said Holmes, laughing.

('The Engineer's Thumb', *Adventures*)

On the more serious side we may remember the story of the criminal Ryder in 'The Blue Carbuncle', a story full of Christmas cheer, but saved from sentimentality by the way Holmes is presented.

He [Ryder] broke into convulsive sobbing, with his face buried in his hands.

There was a long silence, broken only by his heavy breathing, and by the measured tapping of Sherlock Holmes's finger-tips upon the edge of the table. Then my friend rose, and threw open the door.

'Get out!' said he.

'What, sir! Oh, Heaven bless you!'

'No more words. Get out!'

And no more words were needed. There was a rush, a clatter upon the stairs, the bang of a door, and the crisp rattle of running footfalls from the street.

Simone Weil remarks that whereas in real life goodness is wonderful and beautiful, it tends to be rather dull and unconvincing in fiction. This aspect of Sherlock Holmes is surely an exception to that rule, along with Samuel Pickwick and, perhaps, Alyosha Karamazov.

In his preface to the *Case-Book* Doyle placed Holmes 'in the fairy kingdom of romance'. Yet it is precisely that fairy-tale quality that the world of the *Case-Book* lacks. At times we have the sense that we are being given glimpses of hell, as we do with some other work of Doyle: outside the Holmes canon there is 'The Case of Lady Sannox', in which a famous surgeon is tricked into cutting a V-shaped piece out of the lip of his mistress. Nor is 'Lady Sannox' unique in the non-Holmes work. Such stories are reminiscent of Thomas Hardy's story 'Barbara of the House of Grebe', of which T. S. Eliot said that it introduced us into a world of pure Evil. We might also think of 'The Blighting of Sharkey', where a vicious pirate is smitten with a loathsome disease. In the Holmes cycle, 'The Cardboard Box' (*Memoirs*) has its full share of horror, though this is mitigated by the poignancy of the criminal's suffering and by Holmes's compassion. On the whole the world of the *Case-Book* is depicted with a ruthless realism rare in the canon. The stories conjure up a gallery of monstrosity and cruelty:

> the butler and several footmen ran in from the hall. I remember that one of them fainted as I knelt by the injured man and turned that awful face to the light of the lamp. The vitriol was eating into it everywhere and dripping from the ears and the chin. One eye was already white and glazed. The other was red and inflamed. The features which I had admired a few minutes before were now like some beautiful painting over which the artist has passed a wet and foul sponge. They were blurred, discoloured, inhuman, terrible. ('The Illustrious Client')

The horror here is enhanced by the effect of the disfigured face on others ('I remember that one of them fainted').

Similarly, 'Our milkman got a glimpse of her once peeping out of the upper window, and he dropped his tin and the milk all over the front garden' ('The Veiled Lodger'). But it is a particular excellence of this fine story that it gives no description of the shattered face, and the reader sees only 'Two living and beautiful brown eyes looking sadly out from that grisly ruin'.

In 'The Blanched Soldier' the horror is brought into view in a passage whose evocative power must weaken the sceptical view of the story's authenticity: 'The African sun flooded through the big, curtainless windows, and every detail of the great, bare, whitewashed dormitory stood out hard and clear.' Surely the reader is *there*, in that sunlit room. But then comes the plunge into the abnormal.

In front of me was standing a small, dwarf-like man with a huge, bulbous head, who was jabbering excitedly in Dutch, waving two horrible hands which looked to me like brown sponges. Behind him stood a group of people who seemed to be intensely amused by the situation, but a chill came over me as I looked at them. Not one of them was a normal human being. Every one was twisted or swollen or disfigured in some strange way. The laughter of these strange monstrosities was a dreadful thing to hear.

Other Holmes stories, such as 'The Devil's Foot', contain strong meat, 'adult' reading. But in 'The Devil's Foot' the glimpse which Holmes and Watson have of the inner world of the insane is a brief one; they break out of it quite soon; whereas the horror world of the blanched soldier is intensified by its being (as he thinks) permanent.

An effect of the *Case-Book* is Holmes's sudden insight, in 'The Sussex Vampire', into the character of Jacky: 'I watched him', Holmes tells Robert Ferguson, 'as you fondled the child just now. His face was clearly reflected in the glass of the window where the shutter formed a background. I saw such jealousy, such cruel hatred, as I have seldom seen in a human face.' Another horrific picture was forced on Grace Dunbar in 'Thor Bridge': 'When I left her [Mrs Gibson] she was standing still shrieking out her curses at me, in the mouth of the bridge.' In 'The Creeping Man' Professor Presbury's daughter has a similarly intense shock:

As I lay with my eyes fixed upon the square of light, listening to the frenzied barkings of the dog, I was amazed to see my father's face looking in at me. Mr Holmes, I nearly died of surprise and horror. There it was pressed against the window-pane, and one hand seemed to be raised as if to push up the window. If that window had opened, I think I should have gone mad.

Psychological stress is matched in another story by sheer physical pain:

> His back was covered with dark red lines as though he had been terribly flogged by a thin wire scourge. The instrument with which this punishment had been inffiicted was clearly flexible, for the long, angry weals curved round his shoulders and ribs. There was blood dripping down his chin, for he had bitten through his lower lip in the paroxysm of his agony. His drawn and distorted face told how terrible that agony had been. ('The Lion's Mane')

Ugliness is associated with evil. When Holmes reveals that he knows his guilt, the retired colourman 'sprang to his feet with a hoarse scream. He clawed into the air with his bony hands. His mouth was open, and for the instant he looked like some horrible bird of prey.' The world of evil extends itself to the dead: in 'Shoscombe Old Place' Watson says: 'In the glare of the lantern I saw a body swathed in a sheet from head to foot, with dreadful, witch-like features, all nose and chin, projecting at one end, the dim, glazed eyes staring from a discoloured and crumbling face.'

This quality of the *Case-Book* seems to be a blend of Doyle's medical realism with the gruesomeness of Edgar Allan Poe. How did Doyle come to colour his later fiction in this way? Leaving his own psychological problems out of account, much can be attributed to the trauma of the Great War, in which Doyle, like many other men of his age, had lost a son. Wilfred Owen and others denounced the callousness of the old men, but we must not forget David's lament for Absalom, which countless bereaved fathers must have echoed in their thoughts: 'would God I had died for thee, O Absalom, my son, my son!' There was no place for the one-time cosiness of Baker Street in the cruel, disenchanted post-war world.

Such considerations may do something to explain the peculiar timbre of the book. But it is not possible to go any further without confronting what D. Martin Dakin calls 'the problem of the *Case-Book*' (*A Sherlock Holmes Commentary*, 249). Bluntly, it has been held that not only are some of these stories inferior in quality, but they are not the genuine work of Doyle. This 'problem' distinguishes the *Case-Book* from all his other work. Authors are not always at their best, and few would deny that this is reflected in the comparative inferiority of some of the earlier stories. But few have ventured seriously to pronounce them spurious. With the *Case-Book*, on the other hand, there are circumstances which make the suggestion of pseudonymity plausible. We hear of Doyle asking for plots and ideas (he even suggested a competition to find a good plot). It seems clear that stories were sent to him, and he may have adapted them. Boswell said that his own mind was so 'impregnated with the Johnsonian aether' that he could invent convincing conversations in the style of Johnson. Could it be that someone in Doyle's circle—perhaps a member of his family—was similarly impregnated with the Holmesian aether?

We may wonder what would be the result of stylometric analysis of the Holmes stories, the kind of treatment that has been given to the Platonic Dialogues, the plays of Shakespeare, or the Epistles of St Paul. Even without this (and we must remember that such analyses are disputed and controversial) it is hard to believe that any careful reader of the *Case-Book* would be prepared to testify that they all came from the pen of Conan Doyle. One suspicious feature, in the eyes of many readers, is the apparent determination of the Interpolator (if there was one) to exhibit Holmes as a funny man (in the music-hall sense). Examples will be found in 'The Mazarin Stone' (Holmes's banter with Count Negretto Sylvius and Sam Merton), or his exchanges with the black bruiser Steve Dixie and the treacherous cook Susan in 'The Three Gables'. The reader who wishes to apply a touchstone should compare these inferior exchanges with a genuine example of Holmesian humour, taken from *The Valley of Fear*:

> 'Dear me, Watson, is it possible that you have not penetrated the fact that the case hangs upon the missing dumb-bell? Well, well, you need not be downcast, for, between ourselves, I don't think that either Inspector Mac or the excellent local practitioner has grasped the overwhelming importance of this incident. One dumb-bell. Watson! Consider an athlete with one dumb-bell. Picture to yourself the unilateral development, the imminent danger of a spinal curvature. Shocking, Watson; shocking!'

It may be noted that the humour, unlike the Interpolator's facetiousness, is not extraneous but directs attention to the centre of the mystery.

Another suspect feature, which many readers have noticed, is the presence of slang and vulgarisms. Holmes's characteristic mode of speech is dignified; it can support an occasional ascent to rhetorical grandeur; at times, it must be admitted, he verges on the pompous. On the other hand, he can be admirably concise and direct, as when he tells Robert Ferguson: 'Your wife is a very good, a very loving, and a very ill-used woman' ('The Sussex Vampire'). What the genuine Holmes is *not* is vulgar. We suspect an impostor when we hear 'Cut out the poetry, Watson' and 'Pure swank!' in 'The Retired Colourman', and 'His Grace's ma' in 'The Three Gables' (we cannot imagine Holmes talking like that in another tale which touches high life, 'The Priory School'). More serious is the attribution of objectionable attitudes to Holmes; without raising anachronistic questions about Political Correctness, it is hard to accept that a gentleman (which Holmes certainly was) would taunt another man with the colour of his skin (see 'The Three Gables'). Then there is the repetition, in an inferior form, of motifs in the earlier stories. The exchange between Holmes and Lord Bellinger in 'The Second Stain' seems feebly echoed by Holmes's fencing with Colonel Damery over the identity of the Illustrious Client. And the injection of poison into the spaniel in 'The Sussex Vampire' looks like an imitator's re-run of Straker's mutilation of the sheep in 'Silver Blaze'. Finally, though this is hard to substantiate, there is a sub-literary quality in some of the writing. Although Doyle himself professed not to take the Holmes stories very seriously, they are, in their unpretentious way, literature—unlike most of their innumerable imitators. But in 'The Retired Colourman', for example, though it probably contains authentic Holmesian material, we do not feel that we are constantly in touch with the master hand. There is a certain coarseness of style and spirit in this and some of the other tales. But these things do not discredit the whole of the *Case-Book*. 'The Illustrious Client' is as powerful a story as Doyle ever wrote. 'The Sussex Vampire' would not shame any selection of the master's best work. And in the ending of 'The Veiled Lodger' a quiet and unspectacular narrative suddenly achieves towering sublimity.

By way of conclusion, some comments on the *Case-Book* stories in the order of their publication.

'The Mazarin Stone' could serve as an example of the unfortunate influence of the theatre on the stories. It is generally rated as one of the weakest of them. Holmes appears, regrettably, as a variety-show American, with an unworthy opponent. T. S. Eliot illustrated the decline in Doyle's wonderful gift for creating names by contrasting 'Count Negretto Sylvius' with 'Dr Grimesby Roylott of Stoke Moran'. The success of the violin stratagem depends on extraordinary good luck for Holmes and childish folly on the part of the criminal. The trick played on Lord Cantlemere is good. But 'The Mazarin Stone' does not make for serious reading on a mature level. It should be said, however, in view of the plausible hypothesis that Watson is the anonymous author of the tale, that the scenes in which Watson is present are greatly superior to the rest of the story.

'The Problem of Thor Bridge' can surely leave no doubt about its authenticity. The characterization, especially that of Neil Gibson, is thoroughly convincing and the plot is well handled. The inspiration for the solution derived from the work of the German criminologist Hans Gross (1847–1915), whose *System der Kriminalistik* (1891), translated and adapted by John and J. Collyer Adam as *Criminal Investigation* (1907), describes a German suicide who sought to swindle an insurance company by faking his own murder with a pistol tied to a stone which, after the holder had shot himself, dragged the weapon below the water-line beneath the bridge on which the suicide stood: 'the Investigating Officer observed quite by chance that on the decayed wooden parapet of the bridge, almost opposite where the corpse lay, there was a small but perfectly fresh injury which appeared to have been caused by the violent blow on the upper edge of the parapet of a hard and angular body'. He dragged the stream, still believing murder had been committed (for which he had arrested a vagrant), but the pistol, still tied to stone and cord, proved the suicide. Greenhough Smith told Conan Doyle of it, having presumably become interested in Gross's work when he published an article on it (Waldemar Kaempffert, 'The Latest Methods of Tracking Criminals—the "Gross System"', *Strand*, September 1914, and Smith, 'Some Letters of Conan Doyle', *Strand*, October 1930). 'Thor Bridge' has two striking features. One is the Watsonian (or Doylian) stereotype of passionate Hispanic women (see also Mrs Ferguson in 'The Sussex Vampire'). The other is Holmes's stern moral rebuke to the Gold King. This has been objected to as being out of character; but much earlier in the canon we have learned to take Holmes the dispassionate analyst with a pinch of salt: see his treatment of the caddish villain Windibank in 'A Case of Identity' (*Adventures*). It may be noted that Gibson accepts the validity of Holmes's rebuke and manages at the end of it to win a little of our sympathy. Another positive feature of 'Thor Bridge' is our feeling that Holmes is here matched with a strong opponent to be vanquished, before becoming an acceptable client. 'Meaning that I lie,' says Gibson. Holmes replies: 'Well, I was trying to express it as delicately as I could, but if you insist upon the word I will not contradict you.'

'The Creeping Man', if authentic, cannot be read with any great enthusiasm. 'This is another out of the tin box,' says Michael Hardwick, 'which, unfortunately, might as well have been left there. Watson is by turns grumpy, clumsy, and forgetful. The prose shows signs of having been tinkered with, at the least' (*Complete Guide to Sherlock Holmes*, 189). But while 'The Creeping Man' may not be one of the best stories, it is by no means without interest. Professor Presbury, in his aggressive attitude towards visitors, has a faint suggestion of the formidable Professor Challenger (who owed his own survival in *The Lost World* to his ape-like appearance). The Professor's behaviour also has a hint of Mr Hyde's in

Stevenson's *The Strange Case of Dr Jekyll and Mr Hyde* (1886). 'The Creeping Man' is given a further, topical interest by its reflection of the hullabaloo over the Voronoff monkey-gland experiments in the early 1920s. It provides the occasion for Holmes's comment: 'When one tries to rise above Nature one is liable to fall below it.' It is also an opportunity for Holmes's pessimistic thoughts on 'the survival of the least fit': a rebuff to Victorian optimistic evolutionism. A remark of C. S. Lewis can be aptly applied to 'The Creeping Man'. He was expressing disapproval of monkey-gland experiments. His interlocutor protested: 'Oh, come on, Lewis, you'll be an old man yourself one day.' Lewis replied, 'I had rather be an old man than a young monkey.' The authenticity of 'The Creeping Man' is surely proved by Watson's inimitable explanation of the Professor's conduct: 'Lumbago, possibly.'

'The Sussex Vampire' is one of the best of the stories. There can be no reasonable doubt that it is genuine. The information about Watson's early life is a valuable addition to his biography. Holmes's dismissal of the vampire legend is in character, and, if the story was written in the 1920s, it shows great broad-mindedness on the part of Doyle the Spiritualist, for Holmes firmly rules out the supernatural. If the story glances at Bram Stoker's *Dracula*, it is written in a very different spirit. The theme of the Holmes story is anomaly, subtly suggested by the title, in which 'Sussex' and 'Vampire' have contrasting associations (the homely and normal set against the exotic and perverse: a Wellsian juxtaposition). The story can be read as a parable of psychoanalysis, in its aspect as hermeneutics. Ferguson says: 'How can I ever forget how she rose from beside it [the child] with its blood upon her lips?' Holmes shows that there is another way of 'reading' the picture. Holmes is at his best in this story: the aloof physician, but kindly and humane. In spite of (or because of) the strong meat of the story, Holmes's intellectual detachment is marked. 'It has been a case for intellectual deduction, but when this original intellectual deduction is confirmed point by point by quite a number of independent incidents, then the subjective becomes objective and we can say confidently that we have reached our goal.' No wonder 'Ferguson put his big hand to his furrowed forehead'.

'The Three Garridebs' is surely authentic. No lover of the stories would be content to lose the last and best of the Holmes–Watson epiphanies (to use James Joyce's word for them). It occurs when Watson is shot by 'Killer' Evans:

> 'You're not hurt, Watson? For God's sake, say that you are not hurt!'
>
> It was worth a wound—it was worth many wounds—to know the depth of loyalty and love which lay behind that cold mask. The clear, hard eyes were dimmed for a moment, and the firm lips were shaking. For the one and only time I caught a glimpse of a great heart as well as of a great brain. All my years of humble but single-minded service culminated in that moment of revelation.

This, rather than the defeat of Baron Gruner in 'The Illustrious Client', seems the real supreme moment in Holmes's life. The Holmes–Watson relationship is a magnificent example of friendship. (It may be worth pointing out that there is no hint of homosexuality: which cannot be said of the relationship between Raffles and Bunny, created by Doyle's brother-in-law E. W. Hornung.) Apart from this final scene 'The Three Garridebs' is a pleasantly varied *divertimento*. Watson himself has doubts about its genre: 'It may have been a comedy, or it may have been a tragedy. It cost one man his reason, it cost me a blood-letting, and it cost yet another man the penalties of the law. Yet there was certainly an element of comedy.' At any rate, whatever else it is, 'The Three Garridebs' is a masterpiece of story-telling. The plot repeats Doyle's favourite 'decoy' theme.

'The Illustrious Client' is one of the strongest of all the stories, though it is hard to imagine any reader loving it; the characters and incidents are so unattractive. The only jolt in the story is Watson's description of the case as 'the supreme moment of my friend's career'. Watson surely exaggerates here: what about Holmes's services to the nation in such cases as 'The Naval Treaty', 'The Second Stain', or 'The

Bruce-Partington Plans'? This raises the question of Holmes's general attitude to royalty and aristocracy. Where Queen Victoria is concerned Holmes's respect and admiration are as great as Raffles's: a patriotic VR was picked out in bullet-holes on one of the walls of his rooms, and later we hear of the Queen's presenting him with an emerald tie-pin after the affair of the Bruce-Partington plans. But with foreign royalty Holmes's attitude is distinctly non-reverential. In 'A Scandal in Bohemia' we remember Holmes's exchange with the King of Bohemia on the subject of Irene Adler:

King: Is it not a pity that she was not on my level?
Holmes: (coldly) From what I have seen of the lady, she seems, indeed, to have been on a very different level to your Majesty.

Nor is Holmes any more respectful towards home-grown nobility. In 'The Noble Bachelor' Lord Robert St Simon says to him superciliously: 'I understand that you have already managed several delicate cases of this sort, sir, though I presume they were hardly from the same class of society.'

Holmes: No, I am descending.
Lord Robert: I beg pardon?
Holmes: My last client of the sort was a king.

No social taboo inhibits Holmes's stern moral rebuke to the Duke of Holdernesse in his own ducal hall (see 'The Priory School'). In a lighter vein Holmes allows himself a practical joke at the expense of the condescending Lord Cantlemere (see 'The Mazarin Stone'). All these examples, to the democratically minded, are on the credit side. On the debit side we have only Holmes's reference in *The Hound of the Baskervilles* to 'my anxiety to oblige the Pope', which is surely no more than a piece of name-dropping. Outside the canon some have wondered why Doyle did not reprint the powerful non-Holmes 'Story of the Club-Footed Grocer' (*Strand*, November 1898) in book form. Was this because of its low social milieu? He had singled it out as 'a good one' in sending it to Herbert Greenhough Smith back in 1898 (undated MS letter, University of Virginia Library).

But these are trivia in comparison with the crux of 'The Illustrious Client': what appears to be Holmes's obsequiousness towards his client. This is particularly troubling because there is no doubt that the Client is based on Edward VII, and there is evidence both inside and outside the Holmes saga that Doyle's attitude to him both as Prince of Wales and as King was far from uncritical. He is not an attractive figure in the background of 'The Beryl Coronet', pawning priceless Crown property to pay off (presumably) gambling debts or blackmail. And many think that it is he who is disguised as the King of Bohemia. If so, this explains something he has in common with the Illustrious Client: their bizarre notion of disguise. The King of Bohemia appears in flamboyant costume, complete with mask, like someone in R. L. Stevenson's *New Arabian Nights*; Edward VII, for all the fuss about preserving his incognito, drives off in a carriage with his own armorial bearings on it. In the end the question about the Client remains unclear. Doyle (if not Holmes) will have known that, as Martin Dakin put it, the King's interest in young women was by no means always so paternal. Was it because he disapproved of Edward's character that Holmes, we are told, refused a knighthood? On the other hand, Doyle accepted one from Edward. The question, located in the borderlands between reality and fiction, remains murky.

Apart from that, the story has a great deal to be said for it. The snugness of Baker Street has gone; we are in a distinctively Edwardian atmosphere, and have glimpses of a new world, flashy and vulgar. (Is it significant that Watson seems to be avoiding the company of one who belongs to that world, Shinwell Johnson?) Doyle was pleased with 'The Illustrious Client'; he said that, though it was not remarkable for plot, it 'moves adequately in high society'. This last remark is a little puzzling: Is the reference to the kid-gloved Colonel Damery, or to Edward VII? (But we only see his carriage.) Surely it cannot be to Baron Gruner . . . But for all that, many would agree with Doyle's verdict, in his letter to John Gore: 'If I were to choose the six best Holmes stories I should certainly include "The Illustrious Client".' Gruner is the most credible and resourceful of villains. Kitty Winter and Miss de Merville are better drawn and have more verisimilitude than any of the women in the other tales. In their different ways they reinforce a point on which Holmes agrees with Freud: 'Woman's heart and mind are insoluble puzzles to the male.'

'The Three Gables' is a story that many readers would be happy to consider spurious. Holmes's racial gibes at the Negro contrast unpleasantly with the noble anti-racism (approved by both Holmes and Watson) of Grant Munro in 'The Yellow Face'. There are several exasperating exhibitions of Holmes the light comedian, like the banter with Susan ('Paregoric is the stuff'): usually a sign that the Interpolator—if any—is in the vicinity. Another flaw in the story is Holmes's attitude to Isadora Klein, which is excessively arch and indulgent ('he wagged a cautionary forefinger'). This woman had employed 'hired bullies' to maltreat an old lady—even if the offence was to be mitigated by an *amende* of £5,000. Finally, the story is improbable. Someone with Isadora's aplomb should have found it easy to persuade the Duchess of Lomond that she (Mrs Klein) was not portrayed in Douglas Maberley's 'queer novel'. The world depicted in 'The Three Gables' is an unpleasant one. (The Inspector refers to Steve Dixie as 'the big nigger'.) It is a further extension of the world of gangs and beatings-up we were introduced to in 'The Illustrious Client'. Vulgarisms and undistinguished diction are frequent. One hopes that stylistic investigation would leave it in the 'dubious' category (but see the Explanatory Notes for an alternative genesis).

The genuineness of 'The Blanched Soldier' has also been questioned, and this requires some exploration, as more is at stake. Here Holmes himself is the narrator, and we are put off at once by his prolix and fussy manner ('I am compelled to admit that, having taken my pen in my hand,

I do begin to realize that the matter must be presented in such a way as may interest the reader'). Other narratives by Holmes do not lack incisiveness; his contributions to 'The Final Problem' and 'The Empty House' carry the stories vigorously forward, and Holmesian narration makes up a great deal of the substance of 'The *Gloria Scott*' and 'The Musgrave Ritual'. 'The Lion's Mane', another story ostensibly told by Holmes, is not afflicted by the feebleness of 'The Blanched Soldier', although it does not have the bite and vigour of the earlier tales. These two stories have made readers wonder why Holmes could not have found better stories to relate. (And what happened to *The Whole Art of Detection*, which he was supposed to be writing in his retirement?) Finally, as Dakin remarks, the denouement is incredible. In the best of the stories Watson does not contrive such happy endings to please his readers (see, for contrast, 'The Dancing Men' and 'The Greek Interpreter'). It is a formidable indictment. But some of us would, all the same, be reluctant to write off 'The Blanched Soldier'. There is a poignancy about that horribly pale figure, slinking among the shadows of the great house, which goes beyond its puzzle interest. Tuxbury Old Hall, 'a great wandering house, standing in a considerable park . . . all panelling and tapestry and half-effaced old pictures, a house of shadows and mystery' vaguely suggests an inchoate allegory of England in decline. The South African war, which forms a background to the story, foreshadowed the fall of the British Empire. And there is something attractive in the suggestion that in evoking the ghostly presence of young Emsworth, Doyle may have had at the back of his mind the memory of his son Kingsley Conan Doyle, who died of the effects of the 1914–18 war. The poetry, to some extent, atones for its shortcomings. The other memorable quality in the story is the evocation of the leper hospital (quoted earlier), which has the immediacy of direct experience (Doyle had, in 1900, seen the leper island off Cape Town).

'The Lion's Mane' was highly thought of by Doyle himself. He said he 'should put it in the first rank', though he added 'but that is for the public to judge'. However in the *Strand* (March 1927) he said that the story 'is hampered by being told by Holmes himself'. This may be one reason why some readers have judged it to be the weakest of the stories. Like the other purported Holmes narrative, 'The Blanched Soldier', it suffers now and then from a narrative style alien to the true Holmes. And like 'The Blanched Soldier' it makes us conscious of how much is lost by the absence of Watson: tension, mystery, the enhancement of interest by the contrast of excitement with pedestrian sobriety. Moreover the writing, at times admirably plain, can sink to a regrettable flatness. Apart from some fine descriptions of the Sussex coast there is little to put beside the best Watsonian examples of colourful writing—in 'The Final Problem' for example.

A more radical weakness in 'The Lion's Mane' comes out when it is compared with a story which it resembles in structure: 'The Speckled Band' (*Adventures*). In this, too, the agent of evil is a poisonous creature. But what makes 'The Speckled Band' so much more interesting and gripping is that behind the phenomenon of nature is human evil. In the end what defeats the evil of Roylott is human insight and brilliance of intuition, not scientific knowledge. The 'externality' of the story of 'The Lion's Mane' is even more marked in an earlier version that has partially survived: here it is not even Holmes himself, but a famous naturalist, who at the end produces the 'chocolate-backed book' which solves the mystery (we are reminded of the use of Sir James Saunders as a similar *deus ex machina* in 'The Blanched Soldier').

In the best parts of 'The Lion's Mane' Holmesian narrative is indistinguishable from Watsonian: the narrator ceases to be a scientific expert and approximates to a novelist. What some readers remember best about the story is not the Lion's Mane business but the solid and believable characters that it brings together, particularly the girl and the suspected Ian Murdoch. Watson's touch is perceptible here: there is no good reason to question the story's authenticity.

'The Retired Colourman' is a more doubtful case. The opening of the story sounds like unmistakable Holmes, in a mood where the contemplation of the tragic has moved him to philosophical melancholy, as in 'The Boscombe Valley Mystery' (*Adventures*) or the painful story of Browner in 'The Cardboard Box' (*Memoirs*). Otherwise, this story, though containing fine things, is rather rambling, and fails in the end to add up. The uncertainty about the precise status of Barker has troubled some critics, but it does not seem to matter much. The occasional lapses in Holmes's idiom suggest that the Interpolator is again hovering in the background. The chief disappointment in 'The Retired Colourman' is Josiah Amberley himself. The story of how Holmes detects his crime ought to be exciting, but it has the faint mustiness of a report in a *Famous Trials* series. Had Amberley been given a final speech in his true identity at the end it would have united his character, and the story's loose ends, as Browner does so effectively in his confession. But there is no good reason for excluding it from the canon.

'The Veiled Lodger' is on a very different level. This time we have an attractive and strongly sympathetic central figure (Mrs Ronder). The story exhibits a surprising variety of tones. It begins with a delightfully solemn caveat by Watson about attempts that have been made to destroy Holmes's papers, which reveal 'the social and official scandals of the late Victorian era'. This is a reminder that the role of Holmes always involved what Marxist critics like to call a 'contradiction': notoriously the unraveller, the discloser, he is covertly the protector of secrets, the coverer-up. Watson confronts Holmes's anonymous enemies with solemn sternness: 'I deprecate, however, in the strongest way the attempts that have been made lately to get at and destroy these papers. The source of these outrages is known, and if they are repeated I have Mr Holmes's authority for saying that the whole story concerning the politician, the lighthouse, and the trained cormorant will be given to the public. There is at least one reader who will understand.'

But after this promising opening the serio-comic note vanishes from the tale, and we are given instead a detective-story interest, reasonably realistic but without the griminess of 'The Retired Colourman'. The touch is light (at one point the lion who did all the damage makes a brief appearance as a character) and Holmes is not prominent. The sketch of the circus people is attractive (Doyle had evidently read—perhaps with his own children in mind—the circus story for infants in the same issue of the *Strand* (September 1914) as the first instalment of *The Valley of Fear*: see Appendix). The point of the story emerges magnificently in the end, when Holmes has his final exchange with the Veiled Lodger. In view of her ghastly mutilation she suggests she will take her own life. Holmes opposes this with a fine exhortation: 'The example of patient suffering is in itself the most precious of all lessons to an impatient world.' This is as far as charitable humanism can go, by way of consolation. But the woman's answer leaves Holmes with only 'a gesture of pity and protest', and nothing to say:

> She raised her veil and stepped forward into the light.
> 'I wonder if you would bear it', she said.

One thinks—as perhaps Conan Doyle thought—of Tennyson:

> O life as futile, then, as frail!
> O for thy voice to soothe and bless!
> What hope of answer, or redress?
> Behind the veil, behind the veil.
>
> (*In Memoriam*, LVI. vii)

The worst that can be said of 'Shoscombe Old Place' is that it is (for Doyle) rather dull. In the foreground is the life of 'fast' people, gambling, and adultery; in the background, decaying corpses, and the ancient past of England. Sir Robert Norberton turns out to be an unimpressive villain. It is a pity that the last Holmes story should make such anti-climactic use of the motif that has appeared from time to time in the *Case-Book*, the decline of England (for that is what the thematic material of the story works to suggest). The Holmes we have learned to respect and admire, the Victorian Holmes, had been a man at peace with his time; the Holmes of the *Case-Book* is not.

NOTE ON THE TEXT

The text is based on *The Case-Book of Sherlock Holmes* published by John Murray (1927). This has been collated with the individual stories' initial publication in the *Strand* magazine, volumes 62, 63, 65, 67, 69, 72, and 73. The order of appearance in the *Strand* has been followed in preference to the published contents of the Murray collection. The American first publication in *The Case Book of Sherlock Holmes* (*sic*) by George H. Doran, of New York, chronologically simultaneous with the appearance of the Murray text, has been examined for significant variations. 'The Lion's Mane' has been collated against *The Adventure of the Lion's Mane: A fascimile of the original Sherlock Holmes Manuscript with an Introduction by Colin Dexter and Afterword by Richard Lancelyn Green*, its significant variations noted and deleted matter deciphered where possible or conjectured where not. We are profoundly grateful not only to the publishers and custodians of the MS, Westminister Libraries, and their co-publisher, the Sherlock Holmes Society of London, but professionally and personally to Ms Catherine Cooke of the Marylebone Library, in whose Sherlock Holmes Collection it is deposited.

SELECT BIBLIOGRAPHY

1. A. CONAN DOYLE: PRINCIPAL WORKS

(a) *Fiction*

A Study in Scarlet (Ward, Lock, & Co., 1888)
The Mystery of Cloomber (Ward & Downey, 1888)
Micah Clarke (Longmans, Green, & Co., 1889)
The Captain of the Pole-Star and Other Tales (Longmans, Green, & Co., 1890)
The Sign of the Four (Spencer Blackett, 1890)
The Firm of Girdlestone (Chatto & Windus, 1890)
The White Company (Smith, Elder, & Co., 1891)
The Adventures of Sherlock Holmes (George Newnes, 1892)
The Great Shadow (Arrowsmith, 1892)
The Refugees (Longmans, Green, & Co., 1893)
The Memoirs of Sherlock Holmes (George Newnes, 1893)
Round the Red Lamp (Methuen & Co., 1894)
The Stark Munro Letters (Longmans, Green, & Co., 1895)
The Exploits of Brigadier Gerard (George Newnes, 1896)
Rodney Stone (Smith, Elder, & Co., 1896)
Uncle Bernac (Smith, Elder, & Co., 1897)
The Tragedy of the Korosko (Smith, Elder, & Co., 1898)
A Duet With an Occasional Chorus (Grant Richards, 1899)
The Green Flag and Other Stories of War and Sport (Smith, Elder, & Co., 1900)
The Hound of the Baskervilles (George Newnes, 1902)
Adventures of Gerard (George Newnes, 1903)
The Return of Sherlock Holmes (George Newnes, 1905)
Sir Nigel (Smith, Elder, & Co., 1906)
Round the Fire Stories (Smith, Elder, & Co., 1908)
The Last Galley (Smith, Elder, & Co., 1911)
The Lost World (Hodder & Stoughton, 1912)
The Poison Belt (Hodder & Stoughton, 1913)
The Valley of Fear (Smith, Elder, & Co., 1915)
His Last Bow (John Murray, 1917)
Danger! and Other Stories (John Murray, 1918)
The Land of Mist (Hutchinson & Co., 1926)
The Case-Book of Sherlock Holmes (John Murray, 1927)
The Maracot Deep and Other Stories (John Murray, 1929)

The Complete Sherlock Holmes Short Stories (John Murray, 1928)
The Conan Doyle Stories (John Murray, 1929)
The Complete Sherlock Holmes Long Stories (John Murray, 1929)

(b) *Non-fiction*

The Great Boer War (Smith, Elder, & Co., 1900)
The Story of Mr George Edalji (T. Harrison Roberts, 1907)
Through the Magic Door (Smith, Elder, & Co., 1907)
The Crime of the Congo (Hutchinson & Co., 1909)
The Case of Oscar Slater (Hodder & Stoughton, 1912)
The German War (Hodder & Stoughton, 1914)
The British Campaign in France and Flanders (Hodder & Stoughton, 6 vols., 1916–20)
The Poems of Arthur Conan Doyle (John Murray, 1922)
Memories and Adventures (Hodder & Stoughton, 1924; revised ed., 1930)
The History of Spiritualism (Cassell & Co., 1926)

2. MISCELLANEOUS

A Bibliography of A. Conan Doyle (Soho Bibliographies 23: Oxford, 1983) by Richard Lancelyn Green and John Michael Gibson, with a foreword by Graham Greene, is the standard—and indispensable—source of bibliographical information, and of much else besides. Green and Gibson have also assembled and introduced *The Unknown Conan Doyle*, comprising *Uncollected Stories* (those never previously published in book form); *Essays in Photography* (documenting a little-known enthusiasm of Conan Doyle's during his time as a student and young doctor), both published in 1982; and *Letters to the Press* (1986). Alone, Richard Lancelyn Green has compiled (1) *The Uncollected Sherlock Holmes* (1983), an impressive assemblage of Holmesiana, containing almost all Conan Doyle's writing about his creation (other than the stories themselves) together with related material by Joseph Bell, J. M. Barrie, and Beverley Nichols; (2) *The Further Adventures of Sherlock Holmes* (1985), a selection of eleven apocryphal Holmes adventures by various authors, all diplomatically introduced; (3) *The Sherlock Holmes Letters* (1986), a collection of noteworthy public correspondence on Holmes and Holmesiana and far more valuable than its title suggests; and (4) *Letters to Sherlock Holmes* (1984), a powerful testimony to the power of the Holmes stories.

Though much of Conan Doyle's work is now readily available there are still gaps. Some of his very earliest fiction now only survives in rare piracies (apart, that is, from the magazines in which they were first published), including items of intrinsic genre interest such as 'The Gully of Bluemansdyke' (1881) and its sequel 'My Friend the Murderer' (1882), which both turn on the theme of the murderer-informer (handled very differently—and far better—in the Holmes story of 'The Resident Patient' (*Memoirs*)): both of these were used as book-titles for the same pirate collection first issued as *Mysteries and Adventures* (1889). Other stories achieved book publication only after severe pruning—for example, 'The Surgeon of Gaster Fell', reprinted in *Danger!* many years after magazine publication (1890). Some items given initial book publication were not included in the collected edition of *The Conan Doyle Stories.* Particularly deplorable losses were 'John Barrington Cowles' (1884: included subsequently in *Edinburgh Stories of Arthur Conan Doyle* (1981)), 'A Foreign Office Romance' (1894), 'The Club-Footed Grocer' (1898), 'A Shadow Before' (1898), and 'Danger!' (1914). Three of these may have been post-war casualties, as seeming to deal too lightheartedly with the outbreak of other wars; 'John Barrington Cowles' may have been dismissed as juvenile work; but why Conan Doyle discarded a story as good as 'The Club-Footed Grocer' would baffle even Holmes.

At the other end of his life, Conan Doyle's tidying impaired the survival of his most recent work, some of which well merited lasting recognition. *The Maracot Deep and Other Stories* appeared in 1929, a little over a month after *The Conan Doyle Stories*; 'Maracot' itself found a separate paperback life as a short novel; the two Professor Challenger stories, 'The Disintegration Machine' and 'When the World Screamed', were naturally included in John Murray's *The Professor Challenger Stories* (1952); but the fourth item, 'The Story of Spedegue's Dropper', passed beyond the ken of most of Conan Doyle's readers. These three stories show the author, in his seventieth year, still at the height of his powers.

In 1980 Gaslight Publications, of Bloomington, Ind., reprinted *The Mystery of Cloomber*, *The Firm of Girdlestone*, *The Doings of Raffles Haw* (1892), *Beyond the City* (1893), *The Parasite* (1894; also reprinted in *Edinburgh Stories of Arthur Conan Doyle*), *The Stark Munro Letters*, *The Tragedy of the Korosko*, and *A Duet*. *Memories and Adventures*, Conan Doyle's enthralling but impressionistic recollections, are best read in the revised (1930) edition. *Through the Magic Door* remains the best introduction to the literary mind of Conan Doyle, whilst some of his volumes on Spiritualism have autobiographical material of literary significance.

ACD: The Journal of the Arthur Conan Doyle Society (ed. Christopher Roden, David Stuart Davies [to 1991], and Barbara Roden [from 1992]), together with its newsletter, *The Parish Magazine*, is a useful source of critical and biographical material on Conan Doyle. The enormous body of 'Sherlockiana' is best pursued in *The Baker Street Journal*, published by Fordham University Press, or in the *Sherlock Holmes Journal* (Sherlock Holmes Society of London), itemized up to 1974 in the colossal *World Bibliography of Sherlock Holmes and Doctor Watson* (1974) by Ronald Burt De Waal (see also De Waal, *The International Sherlock Holmes* (1980)) and digested in *The Annotated Sherlock Holmes* (2 vols., 1968) by William S. Baring-Gould, whose industry has been invaluable for the Oxford Sherlock Holmes editors. Jack Tracy, *The Encyclopaedia Sherlockiana* (1979) is a very helpful compilation of relevant data. Those who can nerve themselves to consult it despite its title will benefit greatly from Christopher Redmond, *In Bed With Sherlock Holmes* (1984). The classic 'Sherlockian' work is Ronald A. Knox, 'Studies in the Literature of Sherlock Holmes', first published in *The Blue Book* (July 1912) and reprinted in his *Essays in Satire* (1928).

The serious student of Conan Doyle may perhaps deplore the vast extent of 'Sherlockian' literature, even though the size of this output is testimony in itself to the scale and nature of Conan Doyle's achievement. But there is undoubtedly some wheat amongst the chaff. At the head stands Dorothy L. Sayers, *Unpopular Opinions* (1946); also of some interest are T. S. Blakeney, *Sherlock Holmes: Fact or Fiction* (1932), H. W. Bell, *Sherlock Holmes and Dr Watson* (1932), Vincent Starrett, *The Private Life of Sherlock Holmes* (1934), Gavin Brend, *My Dear Holmes* (1951), S. C. Roberts, *Holmes and Watson* (1953) and Roberts's introduction to *Sherlock Holmes: Selected Stories* (Oxford: The World's Classics, 1951), James E. Holroyd, *Baker Street Byways* (1959), Ian McQueen, *Sherlock Holmes Detected* (1974), and Trevor H. Hall, *Sherlock Holmes and his Creator* (1978). One Sherlockian item certainly falls into the category of the genuinely essential: D. Martin Dakin, *A Sherlock Holmes Commentary* (1972), to which all the editors of the present series are indebted.

Michael Pointer, *The Public Life of Sherlock Holmes* (1975) contains invaluable information concerning dramatizations of the Sherlock Holmes stories for radio, stage, and the cinema; of complementary interest are Chris Steinbrunner and Norman Michaels, *The Films of Sherlock Holmes* (1978) and David Stuart Davies, *Holmes of the Movies* (1976), whilst Philip Weller with Christopher Roden, *The Life and Times of Sherlock Holmes* (1992) summarizes a great deal of useful information concerning Conan Doyle's life and Holmes's cases, and in addition is delightfully illustrated. The more concrete products of the Holmes industry are dealt with in Charles Hall, *The Sherlock Holmes Collection* (1987). For a useful retrospective view, Allen Eyles, *Sherlock Holmes: A Centenary Celebration* (1986) rises to the occasion. Both useful and engaging are Peter Haining, *The Sherlock Holmes Scrapbook* (1973) and Charles Viney, *Sherlock Holmes in London* (1989).

Of the many anthologies of Holmesiana, P. A. Shreffler (ed.), *The Baker Street Reader* (1984) is exceptionally useful. D. A. Redmond, *Sherlock Holmes: A Study in Sources* (1982) is similarly indispensable. Michael Hardwick, *The Complete Guide to Sherlock Holmes* (1986) is both reliable and entertaining; Michael Harrison, *In the Footsteps of Sherlock Holmes* (1958) is occasionally helpful.

For more general studies of the detective story, the standard history is Julian Symons, *Bloody Murder* (1972, 1985, 1992). Necessary but a great deal less satisfactory is Howard Haycraft, *Murder for Pleasure* (1942); of more value is Haycraft's critical anthology *The Art of the Mystery Story* (1946), which contains many choice period items. Both R. F. Stewart, *. . . And Always a Detective* (1980) and Colin Watson, *Snobbery with Violence* (1971) are occasionally useful. Dorothy Sayers's pioneering introduction to *Great Short Stories of Detection, Mystery and Horror* (First Series, 1928), despite some inspired howlers, is essential reading; Raymond Chandler's riposte, 'The Simple Art of Murder' (1944), is reprinted in Haycraft, *The Art of the Mystery Story* (see above). Less well known than Sayers's essay but with an equal claim to poineer status is E. M. Wrong's introduction to *Crime and Detection*, First Series (Oxford: The World's Classics, 1926). See also Michael Cox (ed.), *Victorian Tales of Mystery and Detection: An Oxford Anthology* (1992).

Amongst biographical studies of Conan Doyle one of the most distinguished is Jon L. Lellenberg's survey, *The Quest for Sir Arthur Conan Doyle* (1987), with a Foreword by Dame Jean Conan Doyle (much the best piece of writing on ACD by any member of his family). The four earliest biographers—the Revd John Lamond (1931), Hesketh Pearson (1943), John Dickson Carr (1949), and Pierre Nordon (1964)—all had access to the family archives, subsequently closed to researchers following a lawsuit; hence all four biographies contain valuable documentary material, though Nordon handles the evidence best (the French text is fuller than the English version, published in 1966). Of the others, Lamond seems only to have made little use of the material available to him;

Pearson is irreverent and wildly careless with dates; Dickson Carr has a strong fictionalizing element. Both he and Nordon paid a price for their access to the Conan Doyle papers by deferring to the far from impartial editorial demands of Adrian Conan Doyle; Nordon nevertheless remains the best available biography. The best short sketch is Julian Symons, *Conan Doyle* (1979) (and for the late Victorian milieu of the Holmes cycle some of Symons's own fiction, such as *The Blackheath Poisonings* and *The Detling Secret*, can be thoroughly recommended). Harold Orel (ed.), *Critical Essays on Sir Arthur Conan Doyle* (1992) is a good and varied collection, whilst Robin Winks, *The Historian as Detective* (1969) contains many insights and examples applicable to the Holmes corpus; Winks's *Detective Fiction: A Collection of Critical Essays* (1980) is an admirable working handbook, with a useful critical bibliography. Edmund Wilson's famous essay 'Mr Holmes, they were the footprints of a gigantic hound' (1944) may be found in his *Classics and Commercials: A Literary Chronicle of the Forties* (1950).

Specialized biographical areas are covered in Owen Dudley Edwards, *The Quest for Sherlock Holmes: A Biographical Study of Arthur Conan Doyle* (1982) and in Geoffrey Stavert, *A Study in Southsea: The Unrevealed Life of Dr Arthur Conan Doyle* (1987), which respectively assess the significance of the years up to 1882, and from 1882 to 1890. Alvin E. Rodin and Jack D. Key provide a thorough study of Conan Doyle's medical career and its literary implications in *Medical Casebook of Dr Arthur Conan Doyle* (1984). Peter Costello, in *The Real World of Sherlock Holmes: The True Crimes Investigated by Arthur Conan Doyle* (1991) claims too much, but it is useful to be reminded of events that came within Conan Doyle's orbit, even if they are sometimes tangential or even irrelevant. Christopher Redmond, *Welcome to America, Mr Sherlock Holmes* (1987) is a thorough account of Conan Doyle's tour of North America in 1894.

Other than Baring-Gould (see above), the only serious attempt to annotate the nine volumes of the Holmes cycle has been in the Longman Heritage of Literature series (1979–80), to which the present editors are also indebted. Of introductions to individual texts, H. R. F. Keating's to the *Adventures* and *The Hound of the Baskervilles* (published in one volume under the dubious title *The Best of Sherlock Holmes* (1992)) is worthy of particular mention.

A CHRONOLOGY OF ARTHUR CONAN DOYLE

1855 Charles Altamont Doyle, youngest son of the political cartoonist John Doyle ('HB'), and Mary Foley, his Irish landlady's daughter, marry in Edinburgh on 31 July.

1859 Arthur Ignatius Conan Doyle, third child and elder son of ten siblings, born at 11 Picardy Place, Edinburgh, on 22 May and baptized into the Roman Catholic religion of his parents.

1868–75 ACD commences two years' education under the Jesuits at Hodder, followed by five years at its senior sister college, Stonyhurst, both in the Ribble Valley, Lancashire; becomes a popular storyteller amongst his fellow pupils, writes verses, edits a school paper, and makes one close friend, James Ryan of Glasgow and Ceylon. Doyle family resides at 3 Sciennes Hill Place, Edinburgh.

1875–6 ACD passes London Matriculation Examination at Stonyhurst and studies for a year in the Jesuit college at Feldkirch, Austria.

1876–7 ACD becomes a student of medicine at Edinburgh University on the advice of Bryan Charles Waller, now lodging with the Doyle family at 2 Argyle Park Terrace.

1877–80 Waller leases 23 George Square, Edinburgh as a 'consulting pathologist', with all the Doyles as residents. ACD continues medical studies, becoming surgeon's clerk to Joseph Bell at Edinburgh; also takes temporary medical assistantships at Sheffield, Ruyton (Salop), and Birmingham, the last leading to a close friendship with his employer's family, the Hoares. First story published, 'The Mystery of Sasassa Valley', in *Chambers's Journal* (6 Sept. 1879); first non-fiction published—'Gelseminum as a Poison', *British Medical Journal* (20 Sept. 1879). Some time previously ACD sends 'The Haunted Grange of Goresthorpe' to *Blackwood's Edinburgh Magazine*, but it is filed and forgotten.

1880 (Feb.–Sept.) ACD serves as surgeon on the Greenland whaler *Hope* of Peterhead.

1881 ACD graduates MB, CM (Edin.); Waller and the Doyles living at 15 Lonsdale Terrace, Edinburgh.

1881–2 (Oct.–Jan.) ACD serves as surgeon on the steamer *Mayumba* to West Africa, spending three days with US Minister to Liberia, Henry Highland Garnet, black abolitionist leader, then dying. (July–Aug.) Visits Foley relatives in Lismore, Co. Waterford.

1882 Ill-fated partnership with George Turnavine Budd in Plymouth. ACD moves to Southsea, Portsmouth, in June. ACD published in *London Society*, *All the Year Round*, *Lancet*, and *British Journal of Photography*. Over the next eight years ACD becomes an increasingly successful general practitioner at Southsea.

1882–3 Breakup of the Doyle family in Edinburgh. Charles Altamont Doyle henceforth confined because of alcoholism and epilepsy. Mary Foley Doyle resident in Masongill Cottage on the Waller estate at Masongill, Yorkshire. Innes Doyle (b. 1873) resident with ACD as schoolboy and surgery page from Sept. 1882.

1883 'The Captain of the *Pole-Star*' published (*Temple Bar*, Jan.), as well as a steady stream of minor pieces. Works on *The Mystery of Cloomber*.

1884 ACD publishes 'J. Habakuk Jephson's Statement' (*Cornhill Magazine*, Jan.), 'The Heiress of Glenmahowley (*Temple Bar*, Jan.), 'The Cabman's Story' (*Cassell's Saturday Journal*, May); working on *The Firm of Girdlestone*.

1885 Publishes 'The Man from Archangel' (*London Society*, Jan.). Jack Hawkins, briefly a resident patient with ACD, dies of cerebral meningitis. Louisa Hawkins, his sister, marries ACD. (Aug.) Travels in Ireland for honeymoon. Awarded Edinburgh MD.

1886 Writing *A Study in Scarlet*.

1887 *A Study in Scarlet* published in *Beeton's Christmas Annual*.

1888 (July) First book edition of *A Study in Scarlet* published by Ward, Lock; (Dec.) *The Mystery of Cloomber* published.

1889 (Feb.) *Micah Clarke* (ACD's novel of the Monmouth Rebellion of 1685) published. Mary Louise Conan Doyle, ACD's eldest child, born. Unauthorized publication of *Mysteries and Adventures* (published later as *The*

Gully of Bluemansdyke and *My Friend the Murderer*). *The Sign of the Four* and Oscar Wilde's *The Picture of Dorian Gray* commissioned by Lippincott's.

1890 (Jan.) 'Mr [R. L.] Stevenson's Methods in Fiction' published in the *National Review*. (Feb.) *The Sign of the Four* published in *Lippincott's Monthly Magazine*; (Mar.) First authorized short-story collection, *The Captain of the Pole-Star and Other Tales*, published; (Apr.) *The Firm of Girdlestone* published; (Oct.) First book edition of the *Sign* published by Spencer Blackett.

1891 ACD sets up as an eye specialist in 2 Upper Wimpole Street, off Harley Street, while living at Montague Place. Moves to South Norwood. (July–Dec.) The first six 'Adventures of Sherlock Holmes' published in George Newnes's *Strand Magazine*. (Oct.) *The White Company* published; *Beyond the City* first published in *Good Cheer*, the special Christmas number of *Good Words*.

1892 (Jan.–June) Six more Holmes stories published in the *Strand*, with another in Dec. (Mar.) *The Doings of Raffles Haw* published (first serialized in Alfred Harmsworth's penny paper *Answers*, Dec. 1891–Feb. 1892). (14 Oct.) *The Adventures of Sherlock Holmes* published by Newnes. (31 Oct.) Waterloo story *The Great Shadow* published. Alleyne Kingsley Conan Doyle born. Newnes republishes the *Sign*.

1893 'Adventures of Sherlock Holmes' (second series) continues in the *Strand*, to be published by Newnes as *The Memoirs of Sherlock Holmes* (Dec.), minus 'The Cardboard Box'. Holmes apparently killed in 'The Final Problem' (Dec.) to free ACD for 'more serious literary work'. (May) *The Refugees* published. *Jane Annie: or, the Good Conduct Prize* (musical comedy co-written with J. M. Barrie) fails at the Savoy Theatre. (10 Oct.) Charles Altamont Doyle dies.

1894 (Oct.) *Round the Red Lamp*, a collection of medical short stories, published, several for the first time. *The Stark Munro Letters*, a fictionalized autobiography, begun, to be concluded the following year. ACD on US lecture tour with Innes Doyle. (Dec.) *The Parasite* published; 'The Medal of Brigadier Gerard' published in the *Strand*.

1895 'The Exploits of Brigadier Gerard' published in the *Strand*.

1896 (Feb.) *The Exploits of Brigadier Gerard* published by Newnes. ACD settles at Hindhead, Surrey, to minimize effects of his wife's tuberculosis. (Nov.) *Rodney Stone*, a pre-Regency mystery, published. Self-pastiche, 'The Field Bazaar', appears in the Edinburgh University *Student* (20 Nov.).

1897 (May) Napoleonic novel *Uncle Bernac* published; three 'Captain Sharkey' pirate stories published in *Pearson's Magazine* (Jan., Mar., May). Home at Undershaw, Hindhead.

1898 (Feb.) *The Tragedy of the Korosko* published. (June) Publishes *Songs of Action*, a verse collection. (June–Dec.) Begins to publish 'Round the Fire Stories' in the *Strand*—'The Beetle Hunter', 'The Man with the Watches', 'The Lost Special', 'The Sealed Room', 'The Black Doctor', 'The Club-Footed Grocer', and 'The Brazilian Cat'. Ernest William Hornung (ACD's brother-in-law) creates A. J. Raffles and in 1899 dedicates the first stories to ACD.

1899 (Jan.–May) Concludes 'Round the Fire' series in the *Strand* with 'The Japanned Box', 'The Jew's Breast-Plate', 'B. 24', 'The Latin Tutor', and 'The Brown Hand'. (Mar.) Publishes *A Duet with an Occasional Chorus*, a version of his own romance. (Oct.–Dec.) 'The Croxley Master', a boxing story, published in the *Strand*. William Gillette begins 33 years starring in *Sherlock Holmes*, a play by Gillette and ACD.

1900 Accompanies volunteer-staffed Langman hospital as unofficial supervisor to support British forces in the Boer War. (Mar.) Publishes short-story collection, *The Green Flag and Other Stories of War and Sport*. (Oct.) *The Great Boer War* published. Unsuccessful Liberal Unionist parliamentary candidate for Edinburgh Central.

1901 (Aug.) 'The Hound of the Baskervilles' begins serialization in the *Strand*, subtitled 'Another Adventure of Sherlock Holmes'.

1902 (Jan.) *The War in South Africa: Its Cause and Conduct* published. 'Sherlockian' higher criticism begun by Frank Sidgwick in the *Cambridge Review* (23 Jan.). (Mar.) *The Hound of the Baskervilles* published by Newnes. ACD accepts knighthood with reluctance.

1903 (Sept.) *Adventures of Gerard* published by Newnes (previously serialized in the *Strand*). (Oct.) 'The Return of

Sherlock Holmes' begins in the *Strand*. Author's Edition of ACD's major works published in twelve volumes by Smith, Elder and thirteen by D. Appleton & Co. of New York, with prefaces by ACD; many titles omitted.

1904 'Return of Sherlock Holmes' continues in the *Strand*; series designed to conclude with 'The Abbey Grange' (Sept.), but ACD develops earlier allusions and produces 'The Second Stain' (Dec.).

1905 (Mar.) *The Return of Sherlock Holmes* published by Newnes. (Dec.) Serialization of 'Sir Nigel' begun in the *Strand* (concluded Dec. 1906).

1906 (Nov.) Book publication of *Sir Nigel*. ACD defeated as Unionist candidate for Hawick District in general election. (4 July) Death of Louisa ('Touie'), Lady Conan Doyle. ACD deeply affected.

1907 ACD clears the name of George Edalji (convicted in 1903 of cattle-maiming). (18 Sept.) Marries Jean Leckie. (Nov.) Publishes *Through the Magic Door*, a celebration of his literary mentors (earlier version serialized in *Great Thoughts*, 1894).

1908 Moves to Windlesham, Crowborough, Sussex. (Jan.) Death of Sidney Paget. (Sept.) *Round the Fire Stories* published, including some not in earlier *Strand* series. (Sept.–Oct.) 'The Singular Experience of Mr John Scott Eccles' (later retitled as 'The Adventure of Wisteria Lodge') begins occasional series of Holmes stories in the *Strand*.

1909 ACD becomes President of the Divorce Law Reform Union (until 1919). Denis Percy Stewart Conan Doyle born. Takes up agitation against Belgian oppression in the Congo.

1910 (Sept.) 'The Marriage of the Brigadier', the last Gerard story, published in the *Strand*, and (Dec.) the Holmes story of 'The Devil's Foot'. ACD takes six-month lease on Adelphi Theatre; the play *The Speckled Band* opens there, eventually running to 346 performances. Adrian Malcolm Conan Doyle born.

1911 (Apr.) *The Last Galley* (short stories, mostly historical) published. Two more Holmes stories appear in the *Strand*: 'The Red Circle' (Mar., Apr.) and 'The Disappearance of Lady Frances Carfax' (Dec.). ACD declares for Irish Home Rule, under the influence of Sir Roger Casement.

1912 (Apr.–Nov.) The first Professor Challenger story, *The Lost World*, published in the *Strand*, book publication in Oct. Jean Lena Annette Conan Doyle (afterwards Air Commandant Dame Jean Conan Doyle, Lady Bromet) born.

1913 (Feb.) Writes 'Great Britain and the Next War' (*Fortnightly Review*). (Aug.) Second Challenger story, *The Poison Belt*, published. (Dec.) 'The Dying Detective' published in the *Strand*. ACD campaigns for a channel tunnel.

1914 (July) 'Danger!', warning of the dangers of a war-time blockade of Britain, published in the *Strand*. (4 Aug.) Britain declares war on Germany; ACD forms local volunteer force.

1914–15 (Sept.) *The Valley of Fear* begins serialization in the *Strand* (concluding May 1915).

1915 (27 Feb.) *The Valley of Fear* published by George H. Doran in New York. (June) *The Valley of Fear* published in London by Smith, Elder (transferred with rest of ACD stock to John Murray when the firm is sold on the death of Reginald Smith). Five Holmes films released in Germany (ten more during the war).

1916 (Apr., May) First instalments of *The British Campaign in France and Flanders 1914* appear in the *Strand*. (Aug.) *A Visit to Three Fronts* published. Sir Roger Casement convicted of high treason after Dublin Easter Week Rising and executed despite appeals for clemency by ACD and others.

1917 War censor interdicts ACD's history of the 1916 campaigns in the *Strand*. (Sept.) 'His Last Bow' published in the *Strand*. (Oct.) *His Last Bow* published by John Murray (includes 'The Cardboard Box').

1918 (Apr.) ACD publishes *The New Revelation*, proclaiming himself a Spiritualist. (Dec.) *Danger! and Other Stories* published. Permitted to resume accounts of 1916 and 1917 campaigns in the *Strand*, but that for 1918 never serialized. Death of eldest son, Captain Kingsley Conan Doyle, from influenza aggravated by war wounds.

1919 Death of Brigadier-General Innes Doyle, from post-war pneumonia.

1920–30 ACD engaged in world-wide crusade for Spiritualism.

1921–2 ACD's one-act play, *The Crown Diamond*, tours with Dennis Neilson-Terry as Holmes.

1921 (Oct.) 'The Mazarin Stone' (apparently based on *The Crown Diamond*) published in the *Strand*. Death of mother, Mary Foley Doyle.

1922 (Feb.–Mar.) 'The Problem of Thor Bridge' in the *Strand*. (July) John Murray publishes a collected edition of the non-Holmes short stories in six volumes: *Tales of the Ring and the Camp*, *Tales of Pirates and Blue Water*, *Tales of Terror and Mystery*, *Tales of Twilight and the Unseen*, *Tales of Adventure and Medical Life*, and (Nov.) *Tales of Long Ago*. (Sept.) Collected edition of ACD's *Poems* published by Murray.

1923 (Mar.) 'The Creeping Man' published in the *Strand*.

1924 (Jan.) 'The Sussex Vampire' appears in the *Strand*. (June) 'How Watson Learned the Trick', ACD's own Holmes pastiche, appears in *The Book of the Queen's Dolls' House Library*. (Sept.) *Memories and Adventures* published (reprinted with additions and deletions 1930).

1925 (Jan.) 'The Three Garridebs' and (Feb.–Mar.) 'The Illustrious Client' published in the *Strand*. (July) *The Land of Mist*, a Spiritualist novel featuring Challenger, begins serialization in the *Strand*.

1926 (Mar.) *The Land of Mist* published. *Strand* publishes 'The Three Gables' (Oct.), 'The Blanched Soldier' (Nov.), and 'The Lion's Mane' (Dec.).

1927 *Strand* publishes 'The Retired Colourman' (Jan.), 'The Veiled Lodger' (Feb.), and 'Shoscombe Old Place' (Apr.). (June) Murray publishes *The Case-Book of Sherlock Holmes*.

1928 (Oct.) *The Complete Sherlock Holmes Short Stories* published by Murray.

1929 (June) *The Conan Doyle Stories* (containing the six separate volumes issued by Murray in 1922) published. (July) *The Maracot Deep and Other Stories*, ACD's last collection of his fictional work.

1930 (7 July, 8.30 a.m.) Death of Arthur Conan Doyle. 'Education never ends, Watson. It is a series of lessons with the greatest for the last' ('The Red Circle').

The Case-Book of Sherlock Holmes

PREFACE

I FEAR that Mr Sherlock Holmes may become like one of those popular tenors who, having outlived their time, are still tempted to make repeated farewell bows to their indulgent audiences. This must cease and he must go the way of all flesh, material or imaginary. One likes to think that there is some fantastic limbo for the children of imagination, some strange, impossible place where the beaux of Fielding may still make love to the belles of Richardson, where Scott's heroes still may strut, Dickens's delightful Cockneys still raise a laugh, and Thackeray's worldlings continue to carry on their reprehensible careers. Perhaps in some humble corner of such a Valhalla, Sherlock and his Watson may for a time find a place, while some more astute sleuth with some even less astute comrade may fill the stage which they have vacated.

His career has been a long one—though it is possible to exaggerate it; decrepit gentlemen who approach me and declare that his adventures formed the reading of their boyhood do not meet the response from me which they seem to expect. One is not anxious to have one's personal dates handled so unkindly. As a matter of cold fact Holmes made his début in *A Study in Scarlet* and in *The Sign of Four*, two small booklets which appeared between 1887 and 1889 [1890]. It was in 1891 that 'A Scandal in Bohemia', the first of the long series of short stories, appeared in *The Strand Magazine*. The public seemed appreciative and desirous of more, so that from that date, thirty-six years ago, they have been produced in a broken series which now contains no fewer than fifty-six stories, republished in *The Adventures*, *The Memoirs*, *The Return*, and *His Last Bow*, and there remain these twelve published during the last few years which are here produced under the title of *The Case-Book of Sherlock Holmes*. He began his adventures in the very heart of the later Victorian era, carried it through the all-too-short reign of Edward, and has managed to hold his own little niche even in these feverish days. Thus it would be true to say that those who first read of him as young men have lived to see their own grown-up children following the same adventures in the same magazine. It is a striking example of the patience and loyalty of the British public.

I had fully determined at the conclusion of *The Memoirs* to bring Holmes to an end, as I felt that my literary energies should not be directed too much into one channel. That pale, clear-cut face and loose-limbed figure were taking up an undue share of my imagination. I did the deed, but, fortunately, no coroner had pronounced upon the remains, and so, after a long interval, it was not difficult for me to respond to the flattering demand and to explain my rash act away. I have never regretted it, for I have not in actual practice found that these lighter sketches have prevented me from exploring and finding my limitations in such varied branches of literature as history, poetry, historical novels, psychic research, and the drama. Had Holmes never existed I could not have done more, though he may perhaps have stood a little in the way of the recognition of my more serious literary work.

And so, reader, farewell to Sherlock Holmes! I thank you for your past constancy, and can but hope that some return has been made in the shape of that distraction from the worries of life and stimulating change of thought which can only be found in the fairy kingdom of romance.

ARTHUR CONAN DOYLE*

* This was originally published in the *Strand* for March 1927 with slight editorial alteration allowing for 'Shoscombe Old Place' being scheduled for April 1927. It inaugurated a 'Sherlock Holmes Competition' (*Uncollected Sherlock Holmes*, 317–22) and had two other paragraphs relating to that. The *Case-Book* was announced as 'forthcoming' rather than 'here'.

The Mazarin Stone

IT was pleasant to Dr Watson to find himself once more in the untidy room of the first floor in Baker Street which had been the starting-point of so many remarkable adventures. He looked round him at the scientific charts upon the wall, the acid-charred bench of chemicals, the violin-case leaning in the corner, the coal-scuttle, which contained of old the pipes and tobacco. Finally, his eyes came round to the fresh and smiling face of Billy* the young but very wise and tactful page, who had helped a little to fill up the gap of loneliness and isolation which surrounded the saturnine* figure of the great detective.

'It all seems very unchanged, Billy. You don't change, either. I hope the same can be said of him?'*

Billy glanced, with some solicitude, at the closed door of the bedroom.

'I think he's in bed and asleep,' he said.

It was seven in the evening of a lovely summer's day, but Dr Watson was sufficiently familiar with the irregularity of his old friend's hours to feel no surprise at the idea.

'That means a case, I suppose?'

'Yes, sir; he is very hard at it just now. I'm frightened for his health. He gets paler and thinner, and he eats nothing. "When will you be pleased to dine, Mr Holmes?" Mrs Hudson asked. "Seven-thirty, the day after tomorrow," said he. You know his way when he is keen on a case.'

'Yes, Billy, I know.'

'He's following someone. Yesterday he was out as a workman looking for a job. To-day he was an old woman. Fairly took me in, he did, and I ought to know his ways by now.' Billy pointed with a grin to a very baggy parasol which leaned against the sofa. 'That's part of the old woman's outfit,' he said.

'But what is it all about, Billy?'

Billy sank his voice, as one who discusses great secrets of State. 'I don't mind telling you, sir, but it should go no farther. It's this case of the Crown diamond.'

'What—the hundred-thousand-pound burglary?'

'Yes, sir. They must get it back, sir. Why, we had the Prime Minister and the Home Secretary both sitting on that very sofa. Mr Holmes was very nice to them. He soon put them at their ease and promised he would do all he could. Then there is Lord Cantlemere—'*

'Ah!'

'Yes, sir; you know what that means. He's a stiff 'un, sir, if I may say so. I can get along with the Prime Minister, and I've nothing against the Home Secretary, who seemed a civil, obliging sort of man, but I can't stand his lordship. Neither can Mr Holmes, sir. You see, he don't believe in Mr Holmes and he was against employing him. He'd *rather* he failed.'

'And Mr Holmes knows it?'

'Mr Holmes always knows whatever there is to know.'

'Well, we'll hope he won't fail and that Lord Cantlemere will be confounded. But I say, Billy, what is that curtain for across the window?'

'Mr Holmes had it put up there three days ago. We've got something funny behind it.'

Billy advanced and drew away the drapery which screened the alcove of the bow window.

Dr Watson could not restrain a cry of amazement. There was a facsimile* of his old friend, dressing-gown and all, the face turned three-quarters towards the window and downwards, as though reading an invisible book, while the body was sunk deep in an arm-chair. Billy detached the head and held it in the air.

'We put it at different angles, so that it may seem more life-like. I wouldn't dare touch it if the blind were not down. But when it's up you can see this from across the way.'

'We used something of the sort once before.'*

'Before my time,' said Billy. He drew the window curtains apart and looked out into the street. 'There are folk who watch us from over yonder. I can see a fellow now at the window. Have a look for yourself.'

Watson had taken a step forward when the bedroom door opened, and the long, thin form of Holmes emerged, his face pale and drawn, but his step and bearing as active as ever. With a single spring he was at the window, and had drawn the blind once more.

'That will do, Billy,' said he. 'You were in danger of your life then, my boy, and I can't do without you just yet. Well, Watson, it is good to see you in your old quarters once again. You come at a critical moment.'

'So I gather.'

'You can go, Billy. That boy is a problem, Watson. How far am I justified in allowing him to be in danger?'

'Danger of what, Holmes?'

'Of sudden death. I'm expecting something this evening.'

'Expecting what?'

'To be murdered, Watson.'

'No, no; you are joking, Holmes!'

'Even my limited sense of humour could evolve a better joke than that. But we may be comfortable in the mean time, may we not? Is alcohol permitted? The gasogene and cigars are in the old place. Let me see you once more in the customary arm-chair. You have not, I hope, learned to despise my pipe and my lamentable tobacco? It has to take the place of food these days.'

'But why not eat?'

'Because the faculties become refined when you starve them. Why, surely, as a doctor, my dear Watson, you must admit that what your digestion gains in the way of blood supply is so much lost to the brain. I am a brain, Watson. The rest of me is a mere appendix. Therefore, it is the brain I must consider.'

'But this danger, Holmes?'

'Ah, yes; in case it should come off, it would perhaps be as well that you should burden your memory with the name and address of the murderer. You can give it to Scotland Yard, with my love and a parting blessing. Sylvius is the name—Count Negretto Sylvius.* Write it down, man, write it down! 136, Moorside Gardens, NW. Got it?'

Watson's honest face was twitching with anxiety. He knew only too well the immense risks taken by Holmes, and was well aware that what he said was more likely to be under-statement than exaggeration. Watson was always the man of action, and he rose to the occasion.

'Count me in, Holmes. I have nothing to do for a day or two.'

'Your morals don't improve, Watson. You have added fibbing to your other vices. You bear every sign of the busy medical man, with calls on him every hour.'

'Not such important ones. But can't you have this fellow arrested?'

'Yes, Watson, I could. That's what worries him so.'

'But why don't you?'

'Because I don't know where the diamond is.'

'Ah! Billy told me—the missing Crown Jewel!'*

'Yes, the great yellow Mazarin stone.* I've cast my net and I have my fish. But I have not got the stone. What is the use of taking *them*? We can make the world a better place by laying them by the heels. But that is not what I am out for. It's the stone I want.'

'And is this Count Sylvius one of your fish?'

'Yes, and he's a shark. He bites. The other is Sam Merton,* the boxer. Not a bad fellow, Sam, but the Count has used him. Sam's not a shark. He is a great big silly bull-headed gudgeon.* But he is flopping about in my net all the same.'

'Where is this Count Sylvius?'

'I've been at his very elbow all the morning. You've seen me as an old lady, Watson. I was never more convincing. He actually picked up my parasol for me once. "By your leave, madame," said he—half-Italian, you know, and with the Southern graces of manner when in the mood, but a devil incarnate in the other mood. Life is full of whimsical happenings, Watson.'

'It might have been tragedy.'

'Well, perhaps it might. I followed him to old Straubenzee's* workshop in the Minories.* Straubenzee made the air-gun—a very pretty bit of work, as I understand, and I rather fancy it is in the opposite window at the present moment. Have you seen the dummy? Of course, Billy showed it to you. Well, it may get a bullet through its beautiful head at any moment. Ah, Billy, what is it?'

The boy had reappeared in the room with a card upon a tray. Holmes glanced at it with raised eyebrows and an amused smile.

'The man himself. I had hardly expected this. Grasp the nettle, Watson! A man of nerve. Possibly you have heard of his reputation as a shooter of big game. It would indeed be a triumphant ending to his excellent sporting record if he added me to his bag. This is a proof that he feels my toe very close behind his heel.'

'Send for the police.'

'I probably shall. But not just yet. Would you glance carefully out of the window, Watson, and see if anyone is hanging about in the street?'

Watson looked warily round the edge of the curtain.

'Yes, there is one rough fellow near the door.'

'That will be Sam Merton—the faithful but rather fatuous Sam. Where is this gentleman, Billy?'

'In the waiting-room, sir.'*

'Show him up when I ring.'

'Yes, sir.'

'If I am not in the room, show him in all the same.'

'Yes, sir.'

Watson waited until the door was closed, and then he turned earnestly to his companion.

'Look here, Holmes, this is simply impossible. This is a desperate man, who sticks at nothing. He may have come to murder you.'

'I should not be surprised.'

'I insist upon staying with you.'

'You would be horribly in the way.'

'In *his* way?'

'No, my dear fellow—in my way.'

'Well, I can't possibly leave you.'

'Yes, you can, Watson. And you will, for you have never failed to play the game. I am sure you will play it to the end. This man has come for his own purpose, but he may stay for mine.' Holmes took out his notebook and scribbled a few lines. 'Take a cab to Scotland Yard and give this to Youghal of the CID.* Come back with the police. The fellow's arrest will follow.'

'I'll do that with joy.'

'Before you return I may have just time enough to find out where the stone is.' He touched the bell. 'I think we will go out through the bedroom. This second exit is exceedingly useful. I rather want to see my shark without his seeing me, and I have, as you will remember, my own way of doing it.'

It was, therefore, an empty room into which Billy, a minute later, ushered Count Sylvius. The famous game-shot, sportsman, and man-about-town was a big, swarthy fellow, with a formidable dark moustache, shading a cruel, thin-lipped mouth, and surmounted by a long, curved nose, like the beak of an eagle. He was well dressed, but his brilliant necktie, shining pin, and glittering rings were flamboyant in their effect. As the door closed behind him he looked round him with fierce, startled eyes, like one who suspects a trap at every turn. Then he gave a violent start as he saw the impassive head and the collar of the dressing-gown which projected above the arm-chair in the window. At first his expression was one of pure amazement. Then the light of a horrible hope gleamed in his dark, murderous eyes. He took one more glance round to see that there were no witnesses, and then, on tiptoe, his thick stick half raised, he approached the silent figure. He was crouching for his final spring and blow when a cool, sardonic voice greeted him from the open bedroom door:

'Don't break it, Count! Don't break it!'

The assassin staggered back, amazement in his convulsed face. For an instant he half raised his loaded cane once more, as if he would turn his violence from the effigy to the original; but there was something in that steady grey eye and mocking smile which caused his hand to sink to his side.

'It's a pretty little thing,' said Holmes, advancing towards the image. 'Tavernier,* the French modeller made it. He is as good at waxworks as your friend Straubenzee is at air-guns.'

'Air-guns, sir! What do you mean?'

'Put your hat and stick on the side-table. Thank you! Pray take a seat. Would you care to put your revolver out also? Oh, very good, if you prefer to sit upon it. Your visit is really most opportune, for I wanted badly to have a few minutes' chat with you.'

The Count scowled, with heavy, threatening eyebrows.

'I too, wished to have some words with you, Holmes. That is why I am here. I won't deny that I intended to assault you just now.'

Holmes swung his leg on the edge of the table.

'I rather gathered that you had some idea of the sort in your head,' said he. 'But why these personal attentions?'

'Because you have gone out of your way to annoy me. Because you have put your creatures upon my track.'

'My creatures! I assure you no!'

'Nonsense! I have had them followed. Two can play at that game, Holmes.'

'It is a small point, Count Sylvius, but perhaps you would kindly give me my prefix when you address me. You can understand that, with my routine of work, I should find myself on familiar terms with half the rogues' gallery, and you will agree that exceptions are invidious.'

'Well, *Mr* Holmes, then.'

'Excellent! But I assure you you are mistaken about my alleged agents.'

Count Sylvius laughed contemptuously.

'Other people can observe as well as you. Yesterday there was an old sporting man. To-day it was an elderly woman. They held me in view all day.'

'Really, sir, you compliment me. Old Baron Dowson said the night before he was hanged that in my case what the law had gained the stage had lost. And now you give my little impersonations your kindly praise!'

'It was you—you yourself?'

Holmes shrugged his shoulders. 'You can see in the corner the parasol which you so politely handed to me in the Minories before you began to suspect.'

'If I had known, you might never—'

'Have seen this humble home again. I was well aware of it. We all have neglected opportunities to deplore. As it happens, you did not know, so here we are!'

The Count's knotted brows gathered more heavily over his menacing eyes. 'What you say only makes the matter worse. It was not your agents, but your play-acting, busybody self! You admit that you have dogged me. Why?'

'Come now, Count. You used to shoot lions in Algeria.'

'Well?'

'But why?'

'Why? The sport—the excitement—the danger!'

'And, no doubt, to free the country from a pest?'

'Exactly!'

'My reasons in a nutshell!'

The Count sprang to his feet, and his hand involuntarily moved back to his hip-pocket.

'Sit down, sir, sit down! There was another, more practical, reason. I want that yellow diamond!'

Count Sylvius lay back in his chair with an evil smile.

'Upon my word!' said he.

'You knew that I was after you for that. The real reason why you are here to-night is to find out how much I know about the matter and how far my removal is absolutely essential. Well, I should say that, from your point of view, it *is* absolutely essential, for I know all about it, save only one thing, which you are about to tell me.'

'Oh, indeed! And, pray, what is this missing fact?'

'Where the Crown diamond now is.'

The Count looked sharply at his companion. 'Oh, you want to know that, do you? How the devil should I be able to tell you where it is?'

'You can, and you will.'

'Indeed!'

'You can't bluff me, Count Sylvius.' Holmes's eyes, as he gazed at him, contracted and lightened until they were like two menacing points of steel. 'You are absolute plate-glass. I see to the very back of your mind.'

'Then, of course, you see where the diamond is!'

Holmes clapped his hands with amusement, and then pointed a derisive finger. 'Then you do know. You have admitted it!'

'I admit nothing.'

'Now, Count, if you will be reasonable, we can do business. If not, you will get hurt.'

Count Sylvius threw up his eyes to the ceiling. 'And you talk about bluff!' said he.

Holmes looked at him thoughtfully, like a master chess-player who meditates his crowning move. Then he threw open the table drawer and drew out a squat note-book.

'Do you know what I keep in this book?'*

'No, sir, I do not!'

'You!'

'Me?'

'Yes, sir, *you!* You are all here—every action of your vile and dangerous life.'

'Damn you, Holmes!' cried the Count, with blazing eyes. 'There are limits to my patience!'

'It's all here, Count. The real facts as to the death of old Mrs Harold, who left you the Blymer estate, which you so rapidly gambled away.'

'You are dreaming!'

'And the complete life history of Miss Minnie Warrender.'

'Tut! You will make nothing of that!'

'Plenty more here, Count. Here is the robbery in the train-de-luxe* to the Riviera* on February 13, 1892. Here is the forged cheque in the same year on the Credit Lyonnais.'*

'No; you're wrong there.'

'Then I am right on the others! Now, Count, you are a card-player. When the other fellow has all the trumps, it saves time to throw down your hand.'

'What has all this talk to do with the jewel of which you spoke?'

'Gently, Count. Restrain that eager mind! Let me get to the points in my own humdrum fashion. I have all this against you; but, above all, I have a clear case against both you and your fighting bully in the case of the Crown diamond.'

'Indeed!'

'I have the cabman who took you to Whitehall* and the cabman who brought you away. I have the Commissionaire* who saw you near the case. I have Ikey Sanders, who refused to cut it up for you. Ikey has peached,* and the game is up.'

The veins stood out on the Count's forehead. His dark, hairy hands were clenched in a convulsion of restrained emotion. He tried to speak, but the words would not shape themselves.

'That's the hand I play from,' said Holmes. 'I put it all upon the table. But one card is missing. It's the King of Diamonds. I don't know where the stone is.'

'You never shall know.'

'No? Now, be reasonable, Count. Consider the situation. You are going to be locked up for twenty years. So is Sam Merton. What good are you going to get out of your diamond? None in the world. But if you hand it over—well, I'll compound a felony.* We don't want you or Sam. We want the stone. Give that up, and so far as I am concerned you can go free so long as you behave yourself in the future. If you make another slip—well, it will be the last. But this time my commission is to get the stone, not you.'

'But if I refuse?'

'Why, then—alas!—it must be you and not the stone.'

Billy had appeared in answer to a ring.

'I think, Count, that it would be as well to have your friend Sam at this conference. After all, his interests should be represented. Billy, you will see a large and ugly gentleman outside the front door. Ask him to come up.'

'If he won't come, sir?'

'No violence, Billy. Don't be rough with him. If you tell him that Count Sylvius wants him he will certainly come.'

'What are you going to do now?' asked the Count, as Billy disappeared.

'My friend Watson was with me just now. I told him that I had a shark and a gudgeon in my net; now I am drawing the net and up they come together.'

The Count had risen from his chair, and his hand was behind his back. Holmes held something half protruding from the pocket of his dressing-gown.

'You won't die in your bed, Holmes.'

'I have often had the same idea. Does it matter very much? After all, Count, your own exit is more likely to be perpendicular than horizontal. But these anticipations of the future are morbid. Why not give ourselves up to the unrestrained enjoyment of the present?'

A sudden wild-beast light sprang up in the dark, menacing eyes of the master criminal. Holmes's figure seemed to grow taller as he grew tense and ready.

'It is no use your fingering your revolver, my friend,' he said, in a quiet voice. 'You know perfectly well that you dare not use it, even if I gave you time to draw it. Nasty, noisy things, revolvers, Count. Better stick to air-guns. Ah! I think I hear the fairy footstep of your estimable partner. Good day, Mr Merton. Rather dull in the street, is it not?'

The prize-fighter,* a heavily built young man with a stupid, obstinate, slab-sided face, stood awkwardly at the door, looking about him with a puzzled expression. Holmes's debonair manner was a new experience, and though he vaguely felt that it was hostile, he did not know how to counter it. He turned to his more astute comrade for help.

'What's the game now, Count? What's this fellow want? What's up?' His voice was deep and raucous.

The Count shrugged his shoulders and it was Holmes who answered.

'If I may put it in a nutshell, Mr Merton, I should say it was *all* up.'

The boxer still addressed his remarks to his associate.

'Is this cove trying to be funny, or what? I'm not in the funny mood myself.'

'No, I expect not,' said Holmes. 'I think I can promise you that you will feel even less humorous as the evening advances. Now, look here, Count Sylvius. I'm a busy man and I can't waste time. I'm going into that bedroom. Pray make yourselves quite at home in my absence. You can explain to your friend how the matter lies without the restraint of my presence. I shall try over the Hoffmann Barcarolle* upon my violin. In five minutes I shall return for your final answer. You quite grasp the alternative, do you not? Shall we take you, or shall we have the stone?'

Holmes withdrew, picking up his violin from the corner as he passed. A few moments later the long-drawn, wailing notes of that most haunting of tunes came faintly through the closed door of the bedroom.

'What is it, then?' asked Merton anxiously, as his companion turned to him. 'Does he know about the stone?'

'He knows a damned sight too much about it. I'm not sure that he doesn't know all about it.'

'Good Lord!' The boxer's sallow face turned a shade whiter.

'Ikey Sanders has split* on us.'

'He has, has he? I'll do him down a thick 'un* for that if I swing* for it.'

'That won't help us much. We've got to make up our minds what to do.'

'Half a mo',' said the boxer, looking suspiciously at the bedroom door. 'He's a leary* cove that wants watching. I suppose he's not listening?'

'How can he be listening with that music going?'

'That's right. Maybe somebody's behind a curtain. Too many curtains in this room.' As he looked round he suddenly saw for the first time the effigy in the window, and stood staring and pointing, too amazed for words.

'Tut! it's only a dummy,' said the Count.

'A fake, is it? Well, strike me! Madame Tussaud* ain't in it. It's the living spit of him, gown and all. But them curtains, Count!'

'Oh, confound the curtains! We are wasting our time, and there is none too much. He can lag us over this stone.'

'The deuce he can!'

'But he'll let us slip if we only tell him where the swag is.'

'What! Give it up? Give up a hundred thousand quid?'*

'It's one or the other.'

Merton scratched his short-cropped pate.

'He's alone in there. Let's do him in. If his light were out we should have nothing to fear.'

The Count shook his head.

'He is armed and ready. If we shot him we could hardly get away in a place like this. Besides, it's likely enough that the police know whatever evidence he has got. Hallo! What was that?'

There was a vague sound which seemed to come from the window. Both men sprang round, but all was quiet. Save for the one strange figure seated in the chair, the room was certainly empty.

'Something in the street,' said Merton. 'Now look here, guv'nor, you've got the brains. Surely you can think a way out of it. If slugging is no use, then it's up to you.'

'I've fooled better men than he,'* the Count answered. 'The stone is here in my secret pocket. I take no chances leaving it about. It can be out of England to-night and cut into four pieces in Amsterdam* before Sunday. He knows nothing of Van Seddar.'

'I thought Van Seddar was going next week.'

'He *was*. But now he must get off by the next boat. One or other of us must slip round with the stone to Lime Street* and tell him.'

'But the false bottom ain't ready.'

'Well, he must take it as it is and chance it. There's not a moment to lose.' Again, with the sense of danger which becomes an instinct with the sportsman, he paused and looked hard at the window. Yes, it was surely from the street that the faint sound had come.

'As to Holmes,' he continued, 'we can fool him easily enough. You see, the damned fool won't arrest us if he can get the stone. Well, we'll promise him the stone. We'll put him on the wrong track about it, and before he finds that it *is* the wrong track it will be in Holland and we out of the country.'

'That sounds good to me!' said Sam Merton, with a grin.

'You go on and tell the Dutchman to get a move on him. I'll see this sucker and fill him up with a bogus confession. I'll tell him that the stone is in Liverpool. Confound that whining music; it gets on my nerves! By the time he finds it isn't in Liverpool it will be in quarters and we on the blue water. Come back here, out of a line with that keyhole. Here is the stone.'

'I wonder you dare carry it.'

'Where could I have it safer? If we could take it out of Whitehall someone else could surely take it out of my lodgings.'

'Let's have a look at it.'

Count Sylvius cast a somewhat unflattering glance at his associate, and disregarded the unwashed hand which was extended towards him.

'What—d'ye think I'm going to snatch it off you? See here, mister, I'm getting a bit tired of your ways.'

'Well, well; no offence, Sam. We can't afford to quarrel. Come over to the window if you want to see the beauty properly. Now hold it to the light! Here!'

'Thank you!'

With a single spring Holmes had leaped from the dummy's chair and had grasped the precious jewel. He held it now in one hand, while his other pointed a revolver at the Count's head. The two villains staggered back in utter amazement. Before they had recovered Holmes had pressed the electric bell.

'No violence, gentlemen—no violence, I beg of you! Consider the furniture! It must be very clear to you that your position is an impossible one. The police are waiting below.'

The Count's bewilderment overmastered his rage and fear.

'But how the deuce—?' he gasped.

'Your surprise is very natural. You are not aware that a second door* from my bedroom leads behind that curtain. I fancied that you must have heard me when I displaced the figure, but luck was on my side. It gave me a chance of listening to your racy conversation, which would have been painfully constrained had you been aware of my presence.'

The Count gave a gesture of resignation.

'We give you best, Holmes. I believe you are the devil himself.'

'Not far from him, at any rate,' Holmes answered, with a polite smile.

Sam Merton's slow intellect had only gradually appreciated the situation. Now, as the sound of heavy steps came from the stairs outside, he broke silence at last.

'A fair cop!' said he. 'But, I say, what about that bloomin' fiddle! I hear it yet.'

'Tut, tut!' Holmes answered. 'You are perfectly right. Let it play! These modern gramophones* are a remarkable invention.'

There was an inrush of police, the handcuffs clicked, and the criminals were led to the waiting cab. Watson lingered with Holmes, congratulating him upon this fresh leaf added to his laurels. Once more their conversation was interrupted by the imperturbable Billy with his card-tray.

'Lord Cantlemere, sir.'

'Show him up, Billy. This is the eminent peer who represents the very highest interests,' said Holmes. 'He is an excellent and loyal person, but rather of the old *régime*. Shall we make him unbend? Dare we venture upon a slight liberty? He knows, we may conjecture, nothing of what has occurred.'

The door opened to admit a thin, austere figure with a hatchet face and drooping mid-Victorian whiskers* of a glossy blackness* which hardly corresponded with the rounded shoulders and feeble gait. Holmes advanced affably, and shook an unresponsive hand.

'How do you do, Lord Cantlemere? It is chilly, for the time of year, but rather warm indoors. May I take your overcoat?'

'No, I thank you; I will not take it off.'

Holmes laid his hand insistently upon the sleeve.

'Pray allow me! My friend Dr Watson would assure you that these changes of temperature are most insidious.'

His lordship shook himself free with some impatience.

'I am quite comfortable, sir. I have no need to stay. I have simply looked in to know how your self-appointed task was progressing.'

'It is difficult—very difficult.'

'I feared that you would find it so.'

There was a distinct sneer in the old courtier's words and manner.

'Every man finds his limitations, Mr Holmes, but at least it cures us of the weakness of self-satisfaction.'

'Yes, sir, I have been much perplexed.'

'No doubt.'

'Especially upon one point. Possibly you could help me upon it?'

'You apply for my advice rather late in the day. I thought that you had your own all-sufficient methods. Still, I am ready to help you.'

'You see, Lord Cantlemere, we can no doubt frame a case against the actual thieves.'

'When you have caught them.'

'Exactly. But the question is—how shall we proceed against the receiver?'*

'Is this not rather premature?'

'It is as well to have our plans ready. Now, what would you regard as final evidence against the receiver?'

'The actual possession of the stone.'

'You would arrest him upon that?'

'Most undoubtedly.'

Holmes seldom laughed, but he got as near it as his old friend Watson could remember.

'In that case, my dear sir, I shall be under the painful necessity of advising your arrest.'

Lord Cantlemere was very angry. Some of the ancient fires flickered up into his sallow cheeks.

'You take a great liberty, Mr Holmes. In fifty years of official life I cannot recall such a case. I am a busy man, sir, engaged upon important affairs, and I have no time or taste for foolish jokes. I may tell you frankly, sir, that I have never been a believer in your powers, and that I have always been of the opinion that the matter was far safer in the hands of the regular police force. Your conduct confirms all my conclusions. I have the honour, sir, to wish you good evening.'

Holmes had swiftly changed his position and was between the peer and the door.

'One moment, sir,' said he. 'To actually go off with the Mazarin stone would be a more serious offence than to be found in temporary possession of it.'

'Sir, this is intolerable! Let me pass.'

'Put your hand in the right-hand pocket of your overcoat.'

'What do you mean, sir?'

'Come—come; do what I ask.'

An instant later the amazed peer was standing, blinking and stammering, with the great yellow stone on his shaking palm.

'What! What! How is this, Mr Holmes?'

'Too bad, Lord Cantlemere, too bad!' cried Holmes. 'My old friend here will tell you that I have an impish habit of practical joking. Also that I can never resist a dramatic situation. I took the liberty—the very great liberty, I admit—of putting the stone into your pocket at the beginning of our interview.'

The old peer stared from the stone to the smiling face before him.

'Sir, I am bewildered. But—yes—it is indeed the Mazarin stone. We are greatly your debtors, Mr Holmes. Your sense of humour may, as you admit, be somewhat perverted, and its exhibition remarkably untimely, but at least I withdraw any reflection I have made upon your amazing professional powers. But how—'

'The case is but half finished; the details can wait. No doubt, Lord Cantlemere, your pleasure in telling of this successful result in the exalted circle to which you return will be some small atonement for my practical joke. Billy, you will show his lordship out, and tell Mrs Hudson that I should be glad if she would send up dinner for two as soon as possible.'

Thor Bridge

SOMEWHERE in the vaults of the bank of Cox and Co.,* at Charing Cross,* there is a travel-worn and battered tin dispatch-box with my name, John H. Watson, MD, Late Indian Army,* painted upon the lid. It is crammed with papers, nearly all of which are records of cases to illustrate the curious problems which Mr Sherlock Holmes had at various times to examine. Some, and not the least interesting, were complete failures, and as such will hardly bear narrating, since no final explanation is forthcoming. A problem without a solution may interest the student, but can hardly fail to annoy the casual reader. Among these unfinished tales is that of Mr James Phillimore,* who, stepping back into his own house to get his umbrella, was never more seen in this world. No less remarkable is that of the cutter *Alicia** which sailed one spring morning into a small patch of mist from where she never again emerged, nor was anything further ever heard of herself and her crew. A third case worthy of note is that of Isadora Persano,* the well-known journalist and duellist, who was found stark staring mad with a matchbox in front of him which contained a remarkable worm, said to be unknown to science. Apart from these unfathomed cases,* there are some which involve the secrets of private families to an extent which would mean consternation in many exalted quarters if it were thought possible that they might find their way into print. I need not say that such a breach of confidence is unthinkable, and that these records will be separated and destroyed now that my friend has time to turn his energies to the matter. There remain a considerable residue of cases of greater or less interest which I might have edited before had I not feared to give the public a surfeit which might react upon the reputation of the man whom above all others I revere. In some I was myself concerned and can speak as an eye-witness, while in others I was either not present or played so small a part that they could only be told as by a third person.* The following narrative is drawn from my own experience.*

It was a wild morning in October, and I observed as I was dressing how the last remaining leaves were being whirled from the solitary plane tree which graces the yard behind our house. I descended to breakfast prepared to find my companion in depressed spirits, for, like all great artists, he was easily impressed by his surroundings. On the contrary, I found that he had nearly finished his meal, and that his mood was particularly bright and joyous, with that somewhat sinister cheerfulness which was characteristic of his lighter moments.

'You have a case, Holmes?' I remarked.

'The faculty of deduction is certainly contagious, Watson,' he answered. 'It has enabled you to probe my secret. Yes, I have a case. After a month of trivialities and stagnation the wheels move once more.'

'Might I share it?'

'There is little to share, but we may discuss it when you have consumed the two hard-boiled eggs with which our new cook has favoured us. Their condition may not be unconnected with the copy of the *Family Herald** which I observed yesterday upon the hall-table. Even so trivial a matter as cooking an egg demands an attention which is conscious of the passage of time, and incompatible with the love romance in that excellent periodical.'

A quarter of an hour later the table had been cleared and we were face to face. He had drawn a letter from his pocket.

'You have heard of Neil Gibson, the Gold King?' he said.

'You mean the American Senator?'

'Well, he was once Senator for* some Western state, but is better known as the greatest gold-mining magnate* in the world.'

'Yes, I know of him. He has surely lived in England for some time. His name is very familiar.'

'Yes; he bought a considerable estate in Hampshire* some five years ago. Possibly you have already heard of the tragic end* of his wife?'

'Of course. I remember it now. That is why the name is familiar. But I really know nothing of the details.'

Holmes waved his hand towards some papers on a chair. 'I had no idea that the case was coming my way or I should have had my extracts ready,' said he. 'The fact is that the problem, though exceedingly sensational, appeared to present no difficulty. The interesting personality of the accused does not obscure the clearness of the evidence. That was the view taken by the coroner's jury* and also in the police-court proceedings. It is now referred to the Assizes* at Winchester.* I fear it is a thankless business. I can discover facts, Watson, but I cannot change them. Unless some entirely new and unexpected ones come to light I do not see what my client can hope for.'

'Your client?'

'Ah, I forgot I had not told you. I am getting into your involved habit, Watson, of telling a story backwards. You had best read this first.'

The letter which he handed to me, written in a bold, masterful hand, ran as follows:

CLARIDGE'S HOTEL, *October 3rd.*

DEAR MR SHERLOCK HOLMES,

I can't see the best woman God ever made go to her death without doing all that is possible to save her. I can't explain things—I can't even try to explain them, but I know beyond all doubt that Miss Dunbar* is innocent. You know the facts—who doesn't? It has been the gossip of the country. And never a voice raised for her! It's the damned injustice of it all that makes me crazy. That woman has a heart that wouldn't let her kill a fly. Well, I'll come at eleven to-morrow and see if you can get some ray of light in the dark. Maybe I have a clue and don't know it. Anyhow, all I know and all I have and all I am are for your use if only you can save her. If ever in your life you showed your powers, put them now into this case.

Yours faithfully,
J. NEIL GIBSON

'There you have it,' said Sherlock Holmes, knocking out the ashes of his after-breakfast pipe and slowly refilling it. 'That is the gentleman I await. As to the story, you have hardly time to master all these papers, so I must give it to you in a nutshell if you are to take an intellectual interest in the proceedings. This man is the greatest financial power in the world, and a man, as I understand, of most violent and formidable character. He married a wife, the victim of this tragedy, of whom I know nothing save that she was past her prime, which was the more unfortunate as a very attractive governess superintended the education of two young children. These are the three people concerned, and the scene is a grand old manor-house, the centre of an historical English estate. Then as to the tragedy. The wife was found in the grounds nearly half a mile from the house, late at night, clad in her dinner dress, with a shawl over her shoulders and a revolver bullet through her brain. No weapon was found near her and there was no local clue as to the murder. No weapon near her, Watson—mark that! The crime seems to have been committed late in the evening, and the body was found by a gamekeeper about eleven o'clock, when it was examined by the police and by a doctor before being carried up to the house. Is this too condensed, or can you follow it clearly?'

'It is all very clear. But why suspect the governess?'

'Well, in the first place there is some very direct evidence. A revolver with one discharged chamber and a calibre* which corresponded with the bullet was found on the floor of her wardrobe.' His eyes fixed and he repeated in broken words, 'On—the—floor—of—her—wardrobe.' Then he sank into silence, and I saw that some train of thought had been set moving which I should be foolish to interrupt. Suddenly with a start he emerged into brisk life once more. 'Yes, Watson, it was found. Pretty damning, eh? So the two juries* thought. Then the dead woman had a note upon her making an appointment at that very place and signed by the governess. How's that? Finally, there is the motive. Senator Gibson is an attractive person. If his wife dies, who more likely to succeed her than the young lady who had already by all accounts received pressing attentions from her employer. Love, fortune, power, all depending upon one middle-aged life. Ugly, Watson—very ugly!'

'Yes, indeed, Holmes.'

'Nor could she prove an alibi. On the contrary, she had to admit that she was down near Thor Bridge—that was the scene of the tragedy—about that hour. She couldn't deny it, for some passing villager had seen her there.'

'That really seems final.'

'And yet, Watson—and yet! This bridge—a single broad span of stone with balustraded* sides—carries the drive over the narrowest part of a long, deep, reed-girt* sheet of water. Thor Mere* it is called. In the mouth of the bridge lay the dead woman. Such are the main facts. But here, if I mistake not, is our client, considerably before his time.'

Billy* had opened the door, but the name he announced was an unexpected one. Mr Marlow Bates was a stranger to both of us. He was a thin, nervous wisp of a man with frightened eyes, and a twitching, hesitating manner—a man whom my own professional eye would judge to be on the brink of an absolute nervous breakdown.

'You seem agitated, Mr Bates,' said Holmes. 'Pray sit down. I fear I can only give you a short time, for I have an appointment at eleven.'

'I know you have,' our visitor gasped, shooting out short sentences like a man who is out of breath. 'Mr Gibson is coming. Mr Gibson is my employer. I am manager of his estate. Mr Holmes, he is a villain—an infernal villain.'

'Strong language, Mr Bates.'

'I have to be emphatic, Mr Holmes, for the time is so limited. I would not have him find me here for the world. He is almost due now. But I was so situated that I could not come earlier. His secretary, Mr Ferguson, only told me this morning of his appointment with you.'

'And you are his manager?'

'I have given him notice. In a couple of weeks I shall have shaken off his accursed slavery. A hard man, Mr Holmes, hard to all about him. Those public charities are a screen to cover his private iniquities. But his wife was his chief victim. He was brutal to her—yes, sir, brutal! How she came by her death I do not know, but I am sure that he had made her life a misery to her. She was a creature of the Tropics, a Brazilian by birth, as no doubt you know?'

'No; it had escaped me.'

'Tropical by birth and tropical by nature. A child of the sun and of passion. She had loved him as such women can love, but when her own physical charms had faded—I am told that they once were great—there was nothing to hold him. We all liked her and felt for her and hated him for the way that he treated her. But he is plausible and cunning. That is all I have to say to you. Don't take him at his face value. There is more behind. Now I'll go. No, no, don't detain me! He is almost due.'

With a frightened look at the clock our strange visitor literally ran to the door and disappeared.

'Well! well!' said Holmes, after an interval of silence. 'Mr Gibson seems to have a nice loyal household. But the warning is a useful one, and now we can only wait till the man himself appears.'

Sharp at the hour we heard a heavy step upon the stairs and the famous millionaire was shown into the room. As I looked upon him I understood not only the fears and dislike of his manager, but also the execrations which so many business rivals have heaped upon his head. If I were a sculptor and desired to idealize the successful man of affairs, iron of nerve and leathery of conscience, I should choose Mr Neil Gibson as my model. His tall, gaunt, craggy figure had a suggestion of hunger and rapacity. An Abraham Lincoln* keyed to base uses instead of high ones would give some idea of the man. His face might have been chiselled in granite, hard-set, craggy, remorseless, with deep lines upon it, the scars of many a crisis. Cold grey eyes, looking shrewdly out from under bristling brows, surveyed us each in turn. He bowed in perfunctory fashion as Holmes mentioned my name, and then with a masterful air of possession he drew a chair up to my companion and seated himself with his bony knees almost touching him.

'Let me say right here, Mr Holmes,' he began, 'that money is nothing to me in this case. You can burn it if it's any use in lighting you to the truth. This woman is innocent and this woman has to be cleared, and it's up to you to do it. Name your figure!'

'My professional charges are upon a fixed scale,' said Holmes coldly. 'I do not vary them, save when I remit them altogether.'

'Well, if dollars make no difference to you, think of the reputation. If you pull this off every paper* in England and America will be booming you.* You'll be the talk of two continents.'*

'Thank you, Mr Gibson, I do not think that I am in need of booming. It may surprise you to know that I prefer to work anonymously, and that it is the problem itself which attracts me. But we are wasting time. Let us get down to the facts.'

'I think that you will find all the main ones in the Press reports. I don't know that I can add anything which will help you. But if there is anything you would wish more light upon—well, I am here to give it.'

'Well, there is just one point.'

'What is it?'

'What were the exact relations between you and Miss Dunbar?'

The Gold King gave a violent start, and half rose from his chair. Then his massive calm came back to him.

'I suppose you are within your rights—and maybe doing your duty—in asking such a question, Mr Holmes.'

'We will agree to suppose so,' said Holmes.

'Then I can assure you that our relations were entirely and always those of an employer towards a young lady whom he never conversed with, or ever saw, save when she was in the company of his children.'

Holmes rose from his chair.

'I am a rather busy man, Mr Gibson,' said he, 'and I have no time or taste for aimless conversations. I wish you good morning.'

Our visitor had risen also and his great loose figure towered above Holmes. There was an angry gleam from under those bristling brows and a tinge of colour in the sallow cheeks.

'What the devil do you mean by this, Mr Holmes? Do you dismiss my case?'

'Well, Mr Gibson, at least I dismiss you. I should have thought my words were plain.'

'Plain enough, but what's at the back of it? Raising the price on me, or afraid to tackle it, or what? I've a right to a plain answer.'

'Well, perhaps you have,' said Holmes. 'I'll give you one. This case is quite sufficiently complicated to start with, without the further difficulty of false information.'

'Meaning that I lie.'

'Well, I was trying to express it as delicately as I could, but if you insist upon the word I will not contradict you.'

I sprang to my feet, for the expression upon the millionaire's face was fiendish in its intensity, and he had raised his great knotted fist. Holmes smiled languidly, and reached his hand out for his pipe.

'Don't be noisy, Mr Gibson. I find that after breakfast even the smallest argument is unsettling. I suggest that a stroll in the morning air and a little quiet thought will be greatly to your advantage.'

With an effort the Gold King mastered his fury. I could not but admire him, for by a supreme self-command he had turned in a minute from a hot flame of anger to a frigid and contemptuous indifference.

'Well, it's your choice. I guess you know how to run your own business. I can't make you touch the case against your will. You've done yourself no good this morning, Mr Holmes, for I have broken stronger men than you. No man ever crossed me and was the better for it.'

'So many have said so, and yet here I am,' said Holmes, smiling. 'Well, good morning, Mr Gibson. You have a good deal yet to learn.'

Our visitor made a noisy exit, but Holmes smoked in imperturbable silence with dreamy eyes fixed upon the ceiling.

'Any views, Watson?' he asked at last.

'Well, Holmes, I must confess that when I consider that this is a man who would certainly brush any obstacle from his path, and when I remember that his wife may have been an obstacle and an object of dislike, as that man Bates plainly told us, it seems to me—'

'Exactly. And to me also.'

'But what were his relations with the governess, and how did you discover them?'

'Bluff, Watson, bluff! When I considered the passionate, unconventional, unbusiness-like tone of his letter, and contrasted it with his self-contained manner and appearance, it was pretty clear that there was some deep emotion which centred upon the accused woman rather than upon the victim. We've got to understand the exact relations of those three people if we are to reach the truth. You saw the frontal attack which I made upon him and how imperturbably he received it. Then I bluffed him by giving him the impression that I was absolutely certain, when in reality I was only extremely suspicious.'

'Perhaps he will come back?'

'He is sure to come back. He *must* come back. He can't leave it where it is. Ha! isn't that a ring? Yes, there is his footstep. Well, Mr Gibson, I was just saying to Dr Watson that you were somewhat overdue.'

The Gold King had re-entered the room in a more chastened mood than he had left it. His wounded pride still showed in his resentful eyes, but his common sense had shown him that he must yield if he would attain his end.

'I've been thinking it over, Mr Holmes, and I feel that I have been hasty in taking your remarks amiss. You are justified in getting down to the facts, whatever they may be, and I think the more of you for it. I can assure you, however, that the relations between Miss Dunbar and me don't really touch this case.'

'That is for me to decide, is it not?'

'Yes, I guess that is so. You're like a surgeon who wants every symptom before he can give his diagnosis.'

'Exactly. That expresses it. And it is only a patient who has an object in deceiving his surgeon who would conceal the facts of his case.'

'That may be so, but you will admit, Mr Holmes, that most men would shy off a bit when they are asked point-blank what their relations with a woman may be—if there is really some serious feeling in the case. I guess most men have a little private reserve of their own in some corner of their souls where they don't welcome intruders. And you burst suddenly into it. But the object excuses you, since it was to try and save her. Well, the stakes are down and the reserve open and you can explore where you will. What is it you want?'

'The truth.'

The Gold King paused for a moment as one who marshals his thoughts. His grim, deep-lined face had become even sadder and more grave.

'I can give it to you in a very few words, Mr Holmes,' said he at last. 'There are some things that are painful as well as difficult to say, so I won't go deeper than is needful. I met my wife when I was gold-hunting in Brazil. Maria Pinto was the daughter of a Government official at Manaos,* and she was very beautiful. I was young and ardent in those days, but even now, as I look back with colder blood and a more critical eye, I can see that she was rare and wonderful in her beauty. It was a deep rich nature, too, passionate, whole-hearted, tropical, ill-balanced, very different from the American women whom I had known. Well, to make a long story short, I loved her and I married her. It was only when the romance had passed—and it lingered for years—that I realized that we had nothing—absolutely nothing—in common. My love faded. If hers had faded also it might have been easier. But you know the wonderful way of women! Do what I might nothing could turn her from me. If I have been harsh to her, even brutal as some have said, it has been because I knew that if I could kill her love, or if it turned to hate, it would be easier for both of us. But nothing changed her. She adored me in those English woods as she had adored me twenty years ago on the banks of the Amazon.* Do what I might, she was devoted as ever.

'Then came Miss Grace Dunbar. She answered our advertisement and became governess to our two children. Perhaps you have seen her portrait in the papers. The whole world has proclaimed that she also is a very beautiful woman. Now, I make no pretence to be more moral than my neighbours, and I will admit to you that I could not live under the same roof with such a woman and in daily contact with her without feeling a passionate regard for her. Do you blame me, Mr Holmes?'

'I do not blame you for feeling it. I should blame you if you expressed it, since this young lady was in a sense under your protection.'

'Well, maybe so,' said the millionaire, though for a moment the reproof had brought the old angry gleam into his eyes. 'I'm not pretending to be any better than I am. I guess all my life I've been a man that reached out his hand for what he wanted, and I never wanted anything more than the love and possession* of that woman. I told her so.'

'Oh, you did, did you?'

Holmes could look very formidable when he was moved.

'I said to her that if I could marry her I would, but that it was out of my power. I said that money was no object and that all I could do to make her happy and comfortable would be done.'

'Very generous, I am sure,' said Holmes, with a sneer.

'See here, Mr Holmes. I came to you on a question of evidence, not on a question of morals. I'm not asking for your criticism.'

'It is only for the young lady's sake that I touch your case at all,' said Holmes, sternly. 'I don't know that anything she is accused of is really worse than what you have yourself admitted, that you have tried to ruin* a defenceless girl who was under your roof. Some of you rich men have to be taught that all the world cannot be bribed into condoning your offences.'

To my surprise the Gold King took the reproof with equanimity.

'That's how I feel myself about it now. I thank God that my plans did not work out as I intended. She would have none of it, and she wanted to leave the house instantly.'

'Why did she not?'

'Well, in the first place, others were dependent upon her, and it was no light matter for her to let them all down by sacrificing her living. When I had sworn—as I did—that she should never be molested* again, she consented to remain. But there was another reason. She knew the influence she had over me, and that it was stronger than any other influence in the world. She wanted to use it for good.'

'How?'

'Well, she knew something of my affairs. They are large, Mr Holmes—large beyond the belief of an ordinary man. I can make or break—and it is usually break. It wasn't individuals only. It was communities, cities, even nations. Business is a hard game, and the weak go to the wall. I played the game for all it was worth. I never squealed myself and I never cared if the other fellow squealed. But she saw it different. I guess she was right. She believed and said that a fortune for one man that was more than he needed should not be built on ten thousand ruined men who were left without the means of life. That was how she saw it, and I guess she could see past the dollars to something that was more lasting. She found that I listened to what she said, and she believed she was serving the world by influencing my actions. So she stayed—and then this came along.'

'Can you throw any light upon that?'

The Gold King paused for a minute or more, his head sunk in his hands, lost in deep thought.

'It's very black against her. I can't deny that. And women lead an inward life and may do things beyond the judgement of a man. At first I was so rattled and taken aback that I was ready to think she had been led away in some extraordinary fashion that was clean against her usual nature. One explanation came into my head. I give it to you, Mr Holmes, for what it is worth. There is no doubt that my wife was bitterly jealous. There is a soul-jealousy that can be as frantic as any body-jealousy, and though my wife had no cause—and I think she understood this—for the latter,* she was aware that this English girl exerted an influence upon my mind and my acts that she herself never had. It was an influence for good, but that did not mend the matter. She was crazy with hatred, and the heat of the Amazon was always in her blood. She might have planned to murder Miss Dunbar—or we will say to threaten her with a gun and so frighten her into leaving us. Then there might have been a scuffle and the gun gone off and shot the woman who held it.'

'That possibility had already occurred to me,' said Holmes. 'Indeed, it is the only obvious alternative to deliberate murder.'

'But she utterly denies it.'

'Well, that is not final—is it? One can understand that a woman placed in so awful a position might hurry home still in her bewilderment holding the revolver. She might even throw it down among her clothes, hardly knowing what she was doing, and when it was found she might try to lie her way out by a total denial, since all explanation was impossible. What is against such a supposition?'

'Miss Dunbar herself.'

'Well, perhaps.'

Holmes looked at his watch. 'I have no doubt we can get the necessary permits* this morning and reach Winchester by the evening train. When I have seen this young lady, it is very possible that I may be of more use to you in the matter, though I cannot promise that my conclusions will necessarily be such as you desire.'

There was some delay in the official pass, and instead of reaching Winchester that day we went down to Thor Place, the Hampshire estate of Mr Neil Gibson. He did not accompany us himself, but we had the address of Sergeant Coventry, of the local police, who had first examined into the affair. He was a tall, thin, cadaverous man, with a secretive and mysterious manner, which conveyed the idea that he knew or suspected a very great deal more than he dared to say. He had a trick, too, of suddenly sinking his voice to a whisper as if he had come upon something of vital importance, though the information was usually commonplace enough. Behind these tricks of manner he soon showed himself to be a decent, honest fellow who was not too proud to admit that he was out of his depth and would welcome any help.

'Anyhow, I'd rather have you than Scotland Yard, Mr Holmes,' said he. 'If the Yard gets called into a case, then the local loses all credit for success and may be blamed for failure. Now, you play straight, so I've heard.'

'I need not appear in the matter at all,' said Holmes, to the evident relief of our melancholy acquaintance. 'If I can clear it up I don't ask to have my name mentioned.'

'Well, it's very handsome of you, I am sure. And your friend, Dr Watson, can be trusted, I know. Now, Mr Holmes, as we walk down to the place there is one question I should like to ask you. I'd breathe it to no soul but you.' He looked round as though he hardly dare utter the words. 'Don't you think there might be a case against Mr Neil Gibson himself?'

'I have been considering that.'

'You've not seen Miss Dunbar. She is a wonderful fine* woman in every way. He may well have wished his wife out of the road.* And these Americans are readier with pistols than our folk are. It was *his* pistol, you know.'

'Was that clearly made out?'

'Yes, sir. It was one of a pair that he had.'

'One of a pair? Where is the other?'

'Well, the gentleman has a lot of fire-arms of one sort and another. We never quite matched that particular pistol—but the box was made for two.'

'If it was one of a pair you should surely be able to match it.'

'Well, we have them all laid out at the house if you would care to look them over.'

'Later, perhaps. I think we will walk down together and have a look at the scene of the tragedy.'

The conversation had taken place in the little front room of Sergeant Coventry's humble cottage, which served as the local police station. A walk of half a mile or so across a wind-swept heath, all gold and bronze with the fading ferns, brought us to a side-gate opening into the grounds of the Thor Place estate. A path led us through the pheasant preserves,* and then from a clearing we saw the widespread, half-timbered house, half Tudor and half Georgian, upon the crest of the hill. Beside us there was a long, reedy pool, constricted in the centre where the main carriage drive passed over a stone bridge, but swelling into small lakes on either side. Our guide paused at the mouth of this bridge, and he pointed to the ground.

'That was where Mrs Gibson's body lay. I marked it by that stone.'

'I understand that you were there before it was moved?'

'Yes; they sent for me at once.'

'Who did?'

'Mr Gibson himself. The moment the alarm was given and he had rushed down with others from the house, he insisted that nothing should be moved until the police should arrive.'

'That was sensible. I gathered from the newspaper report that the shot was fired from close quarters.'

'Yes, sir, very close.'

'Near the right temple?'

'Just behind it, sir.'

'How did the body lie?'

'On the back, sir. No trace of a struggle. No marks. No weapon. The short note from Miss Dunbar was clutched in her left hand.'

'Clutched, you say?'

'Yes, sir; we could hardly open the fingers.'

'That is of great importance. It excludes the idea that anyone could have placed the note there after death in order to furnish a false clue. Dear me! The note, as I remember, was quite short. "I will be at Thor Bridge at nine o'clock.—G. Dunbar." Was that not so?'

'Yes, sir.'

'Did Miss Dunbar admit writing it?'

'Yes, sir.'

'What was her explanation?'

'Her defence was reserved for the Assizes.* She would say nothing.'

'The problem is certainly a very interesting one. The point of the letter is very obscure, is it not?'

'Well, sir,' said the guide, 'it seemed, if I may be so bold as to say so, the only really clear point in the whole case.'

Holmes shook his head.

'Granting that the letter is genuine and was really written, it was certainly received some time before—say one hour or two. Why, then, was this lady still clasping it in her left hand? Why should she carry it so carefully? She did not need to refer to it in the interview. Does it not seem remarkable?'

'Well, sir, as you put it, perhaps it does.'

'I think I should like to sit quietly for a few minutes and think it out.' He seated himself upon the stone ledge of the bridge, and I could see his quick grey eyes darting their questioning glances in every direction. Suddenly he sprang up again and ran across to the opposite parapet, whipped his lens from his pocket, and began to examine the stonework.

'This is curious,' said he.

'Yes, sir; we saw the chip on the ledge. I expect it's been done by some passer-by.'

The stonework was grey, but at this one point it showed white for a space not larger than a sixpence. When examined closely one could see that the surface was chipped as by a sharp blow.

'It took some violence to do that,' said Holmes thoughtfully. With his cane he struck the ledge several times without leaving a mark. 'Yes, it was a hard knock. In a curious place, too. It was not from above but from below, for you see that it is on the *lower* edge of the parapet.'

'But it is at least fifteen feet from the body.'

'Yes, it is fifteen feet from the body. It may have nothing to do with the matter, but it is a point worth nothing. I do not think we have anything more to learn here. There were no footsteps, you say?'

'The ground was iron hard, sir. There were no traces at all.'

'Then we can go. We will go up to the house first and look over these weapons of which you speak. Then we shall get on to Winchester, for I should desire to see Miss Dunbar before we go farther.'

Mr Neil Gibson had not returned from town, but we saw in the house the neurotic Mr Bates who had called upon us in the morning. He showed us with a sinister relish the formidable array of fire-arms of various shapes and sizes which his employer had accumulated in the course of an adventurous life.

'Mr Gibson has his enemies, as anyone would expect who knew him and his methods,' said he. 'He sleeps with a loaded revolver in the drawer beside his bed. He is a man of violence, sir, and there are times when all of us are afraid of him. I am sure that the poor lady who has passed was often terrified.'

'Did you ever witness physical violence towards her?'

'No, I cannot say that. But I have heard words which were nearly as bad—words of cold, cutting contempt, even before the servants.'

'Our millionaire does not seem to shine in private life,' remarked Holmes, as we made our way to the station. 'Well, Watson, we have come on a good many facts, some of them new ones, and yet I seem some way from my conclusion. In spite of the very evident dislike which Mr Bates has to his employer, I gather from him that when the alarm came he was undoubtedly in his library. Dinner was over at eight-thirty and all was normal up to then. It is true that the alarm was somewhat late in the evening, but the tragedy certainly occurred about the hour named in the note. There is no evidence at all that Mr Gibson had been out of doors since his return from town at five o'clock. On the other hand, Miss Dunbar, as I understand it, admits that she had made an appointment to meet Mrs Gibson at the bridge. Beyond this she would say nothing, as her lawyer had advised her to reserve her defence. We have several very vital questions to ask that young lady, and my mind will not be easy until we have seen her. I must confess that the case would seem to me to be very black against her if it were not for one thing.'

'And what is that, Holmes?'

'The finding of the pistol in her wardrobe.'

'Dear me, Holmes!' I cried, 'that seemed to me to be the most damning incident of all.'

'Not so, Watson. It had struck me even at my first perfunctory reading as very strange, and now that I am in closer touch with the case it is my only firm ground for hope. We must look for consistency. Where there is a want of it we must suspect deception.'

'I hardly follow you.'

'Well, now, Watson, suppose for a moment that we visualize you in the character of a woman who, in a cold, premeditated fashion, is about to get rid of a rival. You have planned it. A note has been written. The victim has come. You have your weapon. The crime is done. It has been workmanlike and complete. Do you tell me that after carrying out so crafty a crime you would now ruin your reputation as a criminal by forgetting to fling your weapon into those adjacent reed-beds which would for ever cover it, but you must needs carry it carefully home and put it in your own wardrobe, the very first place that would be searched? Your best friends would hardly call you a schemer, Watson, and yet I could not picture you doing anything so crude as that.'

'In the excitement of the moment—'

'No, no, Watson, I will not admit that it is possible. Where a crime is coolly premeditated, then the means of covering it are coolly premeditated also. I hope, therefore, that we are in the presence of a serious misconception.'

'But there is so much to explain.'

'Well, we shall set about explaining it. When once your point of view is changed, the very thing which was so damning becomes a clue to the truth. For example, there is the revolver. Miss Dunbar disclaims all knowledge of it. On our new theory she is speaking truth when she says so. Therefore, it was placed in her wardrobe. Who placed it there? Someone who wished to incriminate her. Was not that person the actual criminal? You see how we come at once upon a most fruitful line of inquiry.'

We were compelled to spend the night at Winchester, as the formalities had not yet been completed, but next morning, in the company of Mr Joyce Cummings, the rising barrister* who was entrusted with the defence, we were allowed to see the young lady in her cell. I had expected from all that we had heard to see a beautiful woman, but I can never forget the effect which Miss Dunbar produced upon me. It was no wonder that even the masterful millionaire had found in her something more powerful than himself—something which could control and guide him. One felt, too, as one looked at that strong, clear-cut and yet sensitive face, that even should she be capable of some impetuous deed, none the less there was an innate nobility of character which would make her influence always for the good. She was a brunette, tall, with a noble figure and commanding presence, but her dark eyes had in them this appealing, helpless expression of the hunted creature who feels the nets around it, but can see no way out from the toils. Now, as she realized the presence and the help of my famous friend, there came a touch of colour in her wan cheeks and a light of hope began to glimmer in the glance which she turned upon us.

'Perhaps Mr Neil Gibson has told you something of what occurred between us?' she asked, in a low, agitated voice.

'Yes,' Holmes answered; 'you need not pain yourself by entering into that part of the story. After seeing you,* I am prepared to accept Mr Gibson's statement both as to the influence which you had over him and as to the innocence of your relations with him. But why was the whole situation not brought out in court?'

'It seemed to me incredible that such a charge could be sustained. I thought that if we waited the whole thing must clear itself up without our being compelled to enter into painful details of the inner life of the family. But I understand that far from clearing it has become even more serious.'

'My dear young lady,' cried Holmes, earnestly, 'I beg you to have no illusions upon the point. Mr Cummings here would assure you that all the cards are at present against us, and that we must do everything that is possible if we are to win clear. It would be a cruel deception to pretend that you are not in very great danger. Give me all the help you can, then, to get at the truth.'

'I will conceal nothing.'

'Tell us, then, of your true relations with Mr Gibson's wife.'

'She hated me, Mr Holmes. She hated me with all the fervour of her tropical nature. She was a woman who would do nothing by halves, and the measure of her love for her husband was the measure also of her hatred for me. It is probable that she misunderstood our relations. I would not wish to wrong her, but she loved so vividly in a physical sense that she could hardly understand the mental, and even spiritual, tie which held her husband to me, or imagine that it was only my desire to influence his power to good ends which kept me under his roof. I can see now that I was wrong. Nothing could justify me in remaining where I was a cause of unhappiness, and yet it is certain that the unhappiness would have remained even if I had left the house.'

'Now, Miss Dunbar,' said Holmes, 'I beg you to tell us exactly what occurred that evening.'

'I can tell you the truth so far as I know it, Mr Holmes, but I am in a position to prove nothing, and there are points—the most vital points—which I can neither explain nor can I imagine any explanation.'

'If you will find the facts, perhaps others may find the explanation.'

'With regard, then, to my presence at Thor Bridge that night, I received a note from Mrs Gibson in the morning. It lay on the table of the schoolroom, and it may have been left there by her own hand. It implored me to see her there after dinner, said she had something important to say to me, and asked me to leave an answer on the sundial in the garden, as she desired no one to be in our confidence. I saw no reason for such secrecy, but I did as she asked, accepting the appointment. She asked me to destroy her note and I burned it in the schoolroom grate. She was very much afraid of her husband, who treated her with a harshness for which I frequently reproached him, and I could only imagine that she acted in this way because she did not wish him to know of our interview.'

'Yet she kept your reply very carefully?'

'Yes. I was surprised to hear that she had it in her hand when she died.'

'Well, what happened then?'

'I went down as I had promised. When I reached the bridge she was waiting for me. Never did I realize till that moment how this poor creature hated me. She was like a mad woman—indeed, I think she *was* a mad woman, subtly mad with the deep power of deception which insane people may have. How else could she have met me with unconcern every day and yet had so raging a hatred of me in her heart? I will not say what she said. She poured her whole wild fury out in burning and horrible words. I did not even answer—I could not. It was dreadful to see her. I put my hands to my ears and rushed away. When I left her she was standing still shrieking out her curses at me, in the mouth of the bridge.'

'Where she was afterwards found?'

'Within a few yards from the spot.'

'And yet, presuming that she met her death shortly after you left her, you heard no shot?'

'No, I heard nothing. But, indeed, Mr Holmes, I was so agitated and horrified by this terrible outbreak that I rushed to get back to the peace of my own room, and I was incapable of noticing anything which happened.'

'You say that you returned to your room. Did you leave it again before next morning?'

'Yes; when the alarm came that the poor creature had met her death I ran out with the others.'

'Did you see Mr Gibson?'

'Yes; he had just returned from the bridge when I saw him. He had sent for the doctor and the police.'

'Did he seem to you much perturbed?'

'Mr Gibson is a very strong, self-contained man. I do not think that he would ever show his emotions on the surface. But I, who knew him so well, could see that he was deeply concerned.'

'Then we come to the all-important point. This pistol that was found in your room. Had you ever seen it before?'

'Never, I swear it.'

'When was it found?'

'Next morning, when the police made their search.'

'Among your clothes?'

'Yes; on the floor of my wardrobe under my dresses.'

'You could not guess how long it had been there?'

'It had not been there the morning before.'

'How do you know?'

'Because I tidied out the wardrobe.'

'That is final. Then someone came into your room and placed the pistol there in order to inculpate you.'

'It must have been so.'

'And when?'

'It could only have been at meal-time, or else at the hours when I would be in the schoolroom with the children.'

'As you were when you got the note?'

'Yes; from that time onwards for the whole morning.'

'Thank you, Miss Dunbar. Is there any other point which could help me in the investigation?'

'I can think of none.'

'There was some sign of violence on the stonework of the bridge—a perfectly fresh chip just opposite the body. Could you suggest any possible explanation of that?'

'Surely it must be a mere coincidence.'

'Curious, Miss Dunbar, very curious. Why should it appear at the very time of the tragedy and why at the very place?'

'But what could have caused it? Only great violence could have such an effect.'

Holmes did not answer. His pale, eager face had suddenly assumed that tense, far-away expression which I had learned to associate with the supreme manifestations of his genius. So evident was the crisis in his mind that none of us dared to speak, and we sat, barrister, prisoner, and myself, watching him in a concentrated and absorbed silence. Suddenly he sprang from his chair, vibrating with nervous energy and the pressing need for action.

'Come, Watson, come!' he cried.

'What is it, Mr Holmes?'

'Never mind, my dear lady. You will hear from me, Mr Cummings. With the help of the God of justice* I will give you a case which will make England ring. You will get news by tomorrow, Miss Dunbar, and meanwhile take my assurance that the clouds are lifting and that I have every hope that the light of truth is breaking through.'

It was not a long journey from Winchester to Thor Place, but it was long to me in my impatience, while for Holmes it was evident that it seemed endless; for, in his nervous restlessness, he could not sit still, but paced the carriage or drummed with his long, sensitive fingers upon the cushions beside him.

Suddenly, however, as we neared our destination he seated himself opposite to me—we had a first-class carriage to ourselves—and laying a hand upon each of my knees he looked into my eyes with the peculiarly mischievous gaze which was characteristic of his more imp-like moods.

'Watson,' said he, 'I have some recollection that you go armed upon these excursions of ours.'

It was as well for him that I did so, for he took little care for his own safety when his mind was once absorbed by a problem, so that more than once my revolver had been a good friend in need. I reminded him of the fact.

'Yes, yes, I am a little absent-minded in such matters. But have you your revolver on you?'

I produced it from my hip-pocket, a short, handy, but very serviceable little weapon. He undid the catch, shook out the cartridges, and examined it with care.

'It's heavy—remarkably heavy,' said he.

'Yes, it is a solid bit of work.'

He mused over it for a minute.

'Do you know, Watson,' said he, 'I believe your revolver is going to have a very intimate connection with the mystery which we are investigating.'

'My dear Holmes, you are joking.'

'No, Watson, I am very serious. There is a test before us. If the test comes off, all will be clear. And the test will depend upon the conduct of this little weapon. One cartridge out. Now we will replace the other five and put on the safety-catch.* So! That increases the weight and makes it a better reproduction.'

I had no glimmer of what was in his mind nor did he enlighten me, but sat lost in thought until we pulled up in the little Hampshire station. We secured a ramshackle trap,* and in a quarter of an hour were at the house of our confidential friend, the sergeant.

'A clue, Mr Holmes? What is it?'

'It all depends upon the behaviour of Dr Watson's revolver,' said my friend. 'Here it is. Now, officer, can you give me ten yards of string?'

The village shop provided a ball of stout twine.

'I think that this is all we will need,' said Holmes. 'Now, if you please, we will get off on what I hope is the last stage of our journey.'

The sun was setting and turning the rolling Hampshire moor into a wonderful autumnal panorama. The sergeant, with many critical and incredulous glances, which showed his deep doubts of the sanity of my companion, lurched along beside us. As we approached the scene of the crime I could see that my friend under all his habitual coolness was in truth deeply agitated.

'Yes,' he said, in answer to my remark, 'you have seen me miss my mark before, Watson. I have an instinct for such things, and yet it has sometimes played me false. It seemed a certainty when first it flashed across my mind in the cell at Winchester, but one drawback of an active mind is that one can always conceive alternative explanations which would make our scent a false one. And yet—and yet—Well, Watson, we can but try.'

As he walked he had firmly tied one end of the string to the handle of the revolver. We had now reached the scene of the tragedy. With great care he marked out under the guidance of the policeman the exact spot where the body had been stretched. He then hunted among the heather and the ferns until he found a considerable stone. This he secured to the other end of his line of string, and he hung it over the parapet of the bridge so that it swung clear above the water. He then stood on the fatal spot, some distance from the edge of the bridge, with my revolver in his hand, the string being taut between the weapon and the heavy stone on the farther side.

'Now for it!' he cried.

At the words he raised the pistol to his head, and then let go his grip. In an instant it had been whisked away by the weight of the stone, had struck with a sharp crack against the parapet, and had vanished over the side into the water. It had hardly gone before Holmes was kneeling beside the stonework, and a joyous cry showed that he had found what he expected.

'Was there ever a more exact demonstration?' he cried. 'See, Watson, your revolver has solved the problem!' As he spoke he pointed to a second chip of the exact size and shape of the first which had appeared on the under edge of the stone balustrade.

'We'll stay at the inn to-night,' he continued, as he rose and faced the astonished sergeant. 'You will, of course, get a grappling-hook and you will easily restore my friend's revolver. You will also find beside it the revolver, string and weight with which this vindictive woman attempted to disguise her own crime and to fasten a charge of murder upon an innocent victim. You can let Mr Gibson know that I will see him in the morning, when steps can be taken for Miss Dunbar's vindication.'

Late that evening, as we sat together smoking our pipes in the village inn, Holmes gave me a brief review of what had passed.

'I fear, Watson,' said he, 'that you will not improve any reputation which I may have acquired by adding the Case of the Thor Bridge Mystery to your annals. I have been sluggish in mind and wanting in that mixture of imagination and reality which is the basis of my art. I confess that the chip in the stonework was a sufficient clue to suggest the true solution, and that I blame myself for not having attained it sooner.

'It must be admitted that the workings of this unhappy woman's mind were deep and subtle, so that it was no very simple matter to unravel her plot. I do not think that in our adventures we have ever come across a stranger example of what perverted love can bring about. Whether Miss Dunbar was her rival in a physical or in a merely mental sense seems to have been equally unforgivable in her eyes. No doubt she blamed this innocent lady for all those harsh dealings and unkind words with which her husband tried to repel her too demonstrative affection. Her first resolution was to end her own life. Her second was to do it in such a way as to involve her victim in a fate which was worse far than any sudden death could be.

'We can follow the various steps quite clearly, and they show a remarkable subtlety of mind. A note was extracted very cleverly from Miss Dunbar which would make it appear that she had chosen the scene of the crime. In her anxiety that it should be discovered she somewhat overdid it, by holding it in her hand to the last. This alone should have excited my suspicions earlier than it did.

'Then she took one of her husband's revolvers—there was, as you saw, an arsenal in the house—and kept it for her own use. A similar one she concealed that morning in Miss Dunbar's wardrobe after discharging one barrel, which she could easily do in the woods without attracting attention. She then went down to the bridge where she had contrived this exceedingly ingenious method for getting rid of her weapon. When Miss Dunbar appeared she used her last breath in pouring out her hatred, and then, when she was out of hearing, carried out her terrible purpose. Every link is now in its place and the chain is complete. The papers may ask why the mere was not dragged in the first instance, but it is easy to be wise after the event, and in any case the expanse of a reed-filled lake is no easy matter to drag unless you have a clear perception of what you are looking for and where. Well, Watson, we have helped a remarkable woman, and also a formidable man. Should they in the future join their forces, as seems not unlikely, the financial world may find that Mr Neil Gibson has learned something in that schoolroom of Sorrow where our earthly lessons are taught.'*

The Creeping Man

Mr Sherlock Holmes was always of opinion that I should publish the singular facts connected with Professor Presbury, if only to dispel once for all the ugly rumours which some twenty years ago agitated the University and were echoed in the learned societies of London. There were, however, certain obstacles in the way, and the true history of this curious case remained entombed in the tin box which contains so many records of my friend's adventures. Now we have at last obtained permission to ventilate the facts which formed one of the very last cases handled by Holmes before his retirement from practice. Even now a certain reticence and discretion have to be observed in laying the matter before the public.

It was one Sunday evening early in September of the year 1903 that I received one of the Holmes's laconic messages: 'Come at once if convenient—if inconvenient come all the same.—SH.' The relations between us in those latter days were peculiar. He was a man of habits, narrow and concentrated habits, and I had become one of them. As an institution I was like the violin, the shag tobacco, the old black pipe, the index books, and others perhaps less excusable. When it was a case of active work and a comrade was needed upon whose nerve he could place some reliance, my *rôle* was obvious. But apart from this I had uses. I was a whetstone for his mind. I stimulated him. He liked to think aloud in my presence. His remarks could hardly be said to be made to me—many of them would have been as appropriately addressed to his bedstead—but none the less, having formed the habit, it had become in some way helpful that I should register and interject. If I irritated him by a certain methodical slowness in my mentality, that irritation served only to make his own flame-like intuitions and impressions flash up the more vividly and swiftly. Such was my humble *rôle* in our alliance.

When I arrived at Baker Street* I found him huddled up in his arm-chair with updrawn knees, his pipe in his mouth and his brow furrowed with thought. It was clear that he was in the throes of some vexatious problem. With a wave of his hand he indicated my old arm-chair, but otherwise for half an hour he gave no sign that he was aware of my presence. Then with a start he seemed to come from his reverie, and, with his usual whimsical smile, he greeted me back to what had once been my home.

'You will excuse a certain abstraction of mind, my dear Watson,' said he. 'Some curious facts have been submitted to me within the last twenty-four hours, and they in turn have given rise to some speculations of a more general character. I have serious thoughts of writing a small monograph* upon the uses of dogs in the work of the detective.'

'But surely, Holmes, this has been explored,' said I. 'Bloodhounds—sleuth-hounds—'

'No, no, Watson; that side of the matter is, of course, obvious. But there is another which is far more subtle. You may recollect that in the case which you, in your sensational way, coupled with the Copper Beeches,* I was able, by watching the mind of the child, to form a deduction as to the criminal habits of the very smug and respectable father.'

'Yes, I remember it well.'

'My line of thoughts about dogs is analogous. A dog reflects the family life. Whoever saw a frisky dog in a gloomy family, or a sad dog in a happy one? Snarling people have snarling dogs, dangerous people have dangerous ones. And their passing moods may reflect the passing moods of others.'

I shook my head. 'Surely, Holmes, this is a little far-fetched,' said I.

He had refilled his pipe and resumed his seat, taking no notice of my comment.

'The practical application of what I have said is very close to the problem which I am investigating. It is a tangled skein,* you understand, and I am looking for a loose end. One possible loose end lies in the question: Why does Professor Presbury's faithful wolf-hound, Roy,* endeavour to bite him?'

I sank back in my chair in some disappointment. Was it for so trivial a question as this that I had been summoned from my work? Holmes glanced across at me.

'The same old Watson!' said he. 'You never learn that the gravest issues may depend upon the smallest things. But is it not on the face of it strange that a staid, elderly philosopher—you've heard of Presbury, of course, the famous Camford physiologist?*—that such a man, whose friend has been his devoted wolf-hound, should now have been twice attacked by his own dog? What do you make of it?'

'The dog is ill.'

'Well, that has to be considered. But he attacks no one else, nor does he apparently molest his master, save on very special occasions. Curious, Watson—very curious. But young Mr Bennett is before his time, if that is his ring. I had hoped to have a longer chat with you before he came.'

There was a quick step on the stairs, a sharp tap at the door, and a moment later the new client presented himself. He was a tall, handsome youth about thirty, well dressed and elegant, but with something in his bearing which suggested the shyness of the student rather than the self-possession of the man of the world. He shook hands with Holmes, and then looked with some surprise at me.

'This matter is very delicate, Mr Holmes,' he said. 'Consider the relation in which I stand to Professor Presbury, both privately and publicly. I really can hardly justify myself if I speak before any third person.'

'Have no fear, Mr Bennett. Dr Watson is the very soul of discretion, and I can assure you that this is a matter in which I am very likely to need an assistant.'

'As you like, Mr Holmes. You will, I am sure, understand my having some reserves in the matter.'

'You will appreciate it, Watson, when I tell you that this gentleman, Mr Trevor Bennett, is professional assistant to the great scientist, lives under his roof, and is engaged to his only daughter. Certainly we must agree that the Professor has every claim upon his loyalty and devotion. But it may best be shown by taking the necessary steps to clear up this strange mystery.'

'I hope so, Mr Holmes. That is my one object. Does Dr Watson know the situation?'

'I have not had time to explain it.'

'Then perhaps I had better go over the ground again before explaining some fresh developments.'

'I will do so myself,' said Holmes, 'in order to show that I have the events in their due order. The Professor, Watson, is a man of European reputation. His life has been academic. There has never been a breath of scandal. He is a widower with one daughter, Edith. He is, I gather, a man of very virile and positive, one might almost say combative, character. So the matter stood until a very few months ago.

'Then the current of his life was broken. He is sixty-one years of age, but he became engaged to the daughter of Professor Morphy, his colleague in the chair of Comparative Anatomy.* It was not, as I understand, the reasoned courting of an elderly man, but rather the passionate frenzy of youth, for no one could have shown himself a more devoted lover. The lady, Alice Morphy, was a very perfect girl both in mind and body, so that there was every excuse for the Professor's infatuation. None the less, it did not meet with full approval in his own family.'

'We thought it rather excessive,' said our visitor.

'Exactly. Excessive and a little violent and unnatural. Professor Presbury was rich, however, and there was no objection upon the part of the father. The daughter, however, had other views, and there were already several candidates for her hand, who, if they were less eligible from a worldly point of view, were at least more of an age. The girl seemed to like the Professor in spite of his eccentricities. It was only age which stood in the way.

'About this time a little mystery suddenly clouded the normal routine of the Professor's life. He did what he had never done before. He left home and gave no indication where he was going. He was away a fortnight, and returned looking rather travel-worn. He made no allusion to where he had been, although he was usually the frankest of men. It chanced, however, that our client here, Mr Bennett, received a letter from a fellow-student in Prague, who said that he was glad to have seen Professor Presbury there, although he had not been able to talk to him. Only in this way did his own household learn where he had been.

'Now comes the point. From that time onwards a curious change came over the Professor. He became furtive and sly. Those around him had always the feeling that he was not the man that they had known, but that he was under some shadow which had darkened his higher qualities. His intellect was not affected. His lectures were as brilliant as ever. But always there was something new, something sinister and unexpected. His daughter, who was devoted to him, tried again and again to resume the old relations and to penetrate this mask which her father seemed to have put on. You, sir, as I understand, did the same—but all was in vain. And now, Mr Bennett, tell in your own words the incident of the letters.'

'You must understand, Dr Watson, that the Professor had no secrets from me. If I were his son or his younger brother, I could not have more completely enjoyed his confidence. As his secretary I handled every paper which came to him, and I opened and subdivided his letters. Shortly after his return all this was changed. He told me that certain letters might come to him from London which would be marked by a cross under the stamp. These were to be set aside for his own eyes only. I may say that several of these did pass through my hands, that they had the E.C. mark,* and were in an illiterate handwriting. If he answered them at all the answers did not pass through my hands nor into the letter-basket in which our correspondence was collected.'

'And the box,' said Holmes.

'Ah, yes, the box. The Professor brought back a little wooden box from his travels. It was the one thing which suggested a Continental tour, for it was one of those quaint carved things which one associates with Germany. This he placed in his instrument cupboard. One day, in looking for a cannula,* I took up the box. To my surprise he was very angry, and reproved me in words which were quite savage for my curiosity. It was the first time such a thing had happened and I was deeply hurt. I endeavoured to explain that it was a mere accident that I had touched the box, but all the evening I was conscious that he looked at me harshly and that the incident was rankling in his mind.' Mr Bennett drew a little diary book from his pocket. 'That was on July 2nd,' said he.

'You are certainly an admirable witness,' said Holmes. 'I may need some of these dates which you have noted.'

'I learned method among other things from my great teacher. From the time that I observed abnormality in his behaviour I felt that it was my duty to study his case. Thus I have it here that it was on that very day, July 2nd, that Roy attacked the Professor as he came from his study into the hall. Again on July 11th there was a scene of the same sort, and then I have a note of yet another upon July 20th. After that we had to banish Roy to the stables. He was a dear, affectionate animal—but I fear I weary you.'

Mr Bennett spoke in a tone of reproach, for it was very clear that Holmes was not listening. His face was rigid and his eyes gazed abstractedly at the ceiling. With an effort he recovered himself.

'Singular! Most singular!' he murmured. 'These details were new to me, Mr Bennett. I think we have now fairly gone over the old ground, have we not? But you spoke of some fresh developments.'

The pleasant, open face of our visitor clouded over, shadowed by some grim remembrance. 'What I speak of occurred the night before last,' said he. 'I was lying awake about two in the morning, when I was aware of a dull muffled sound coming from the passage. I opened my door and peeped out. I should explain that the Professor sleeps at the end of the passage—'

'The date being—?' asked Holmes.

Our visitor was clearly annoyed at so irrelevant an interruption.

'I have said, sir, that it was the night before last—that is, September 4th.'

Holmes nodded and smiled.

'Pray continue,' said he.

'He sleeps at the end of the passage, and would have to pass my door in order to reach the staircase. It was a really terrifying experience, Mr Holmes. I think that I am as strong-nerved as my neighbours, but I was shaken by what I saw. The passage was dark save that one window half-way along it threw a patch of light. I could see that something was coming along the passage, something dark and crouching. Then suddenly it emerged into the light, and I saw that it was he. He was crawling, Mr Holmes—crawling! He was not quite on his hands and knees. I should rather say on his hands and feet, with his face sunk between his hands. Yet he seemed to move with ease. I was so paralysed by the sight that it was not until he had reached my door that I was able to step forward and ask if I could assist him. His answer was extraordinary. He sprang up, spat out some atrocious word at me, and hurried on past me, and down the staircase. I waited about for an hour, but he did not come back. It must have been daylight before he regained his room.'

'Well, Watson, what make you of that?' asked Holmes, with the air of the pathologist* who presents a rare specimen.

'Lumbago,* possibly. I have known a severe attack make a man walk in just such a way, and nothing would be more trying to the temper.'

'Good, Watson! You always keep us flat-footed on the ground. But we can hardly accept lumbago, since he was able to stand erect in a moment.'

'He was never better in health,' said Bennett. 'In fact, he is stronger than I have known him for years. But there are the facts, Mr Holmes. It is not a case in which we can consult the police, and yet we are utterly at our wits' end as to what to do, and we feel in some strange way that we are drifting towards disaster. Edith—Miss Presbury—feels as I do, that we cannot wait passively any longer.'

'It is certainly a very curious and suggestive case. What do you think, Watson?'

'Speaking as a medical man,' said I, 'it appears to be a case for an alienist.* The old gentleman's cerebral processes were disturbed by the love affair. He made a journey abroad in the hope of breaking himself of the passion. His letters and the box may be connected with some other private transaction—a loan, perhaps, or share certificates,* which are in the box.'

'And the wolf-hound no doubt disapproved of the financial bargain. No, no, Watson, there is more in it than this. Now, I can only suggest—'

What Sherlock Holmes was about to suggest will never be known, for at this moment the door opened and a young lady was shown into the room. As she appeared Mr Bennett sprang up with a cry and ran forward with his hands out to meet those which she had herself outstretched.

'Edith, dear! Nothing the matter, I hope?'

'I felt I must follow you. Oh, Jack,* I have been so dreadfully frightened! It is awful to be there alone.'

'Mr Holmes, this is the young lady I spoke of. This is my *fiancée*.'

'We were gradually coming to that conclusion, were we not, Watson?' Holmes answered, with a smile. 'I take it, Miss Presbury, that there is some fresh development in the case, and that you thought we should know?'

Our new visitor, a bright, handsome girl of a conventional English type, smiled back at Holmes as she seated herself beside Mr Bennett.

'When I found Mr Bennett had left his hotel I thought I should probably find him here. Of course, he had told me that he would consult you. But, oh, Mr Holmes, can you do nothing for my poor father?'

'I have hopes, Miss Presbury, but the case is still obscure. Perhaps what you have to say may throw some fresh light upon it.'

'It was last night, Mr Holmes. He had been very strange all day. I am sure that there are times when he has no recollection of what he does. He lives as in a strange dream. Yesterday was such a day. It was not my father with whom I lived. His outward shell was there, but it was not really he.'

'Tell me what happened.'

'I was awakened in the night by the dog barking most furiously. Poor Roy, he is chained now near the stable. I may say that I always sleep with my door locked; for, as Jack—as Mr Bennett—will tell you, we all have a feeling of impending danger. My room is on the second floor. It happened that the blind was up in my window, and there was bright moonlight outside. As I lay with my eyes fixed upon the square of light, listening to the frenzied barkings of the dog, I was amazed to see my father's face looking in at me. Mr Holmes, I nearly died of surprise and horror. There it was pressed against the window-pane, and one hand seemed to be raised as if to push up the window. If that window had opened, I think I should have gone mad. It was no delusion, Mr Holmes. Don't deceive yourself by thinking so. I dare say it was twenty seconds or so that I lay paralysed and watched the face. Then it vanished, but I could not—I could not spring out of bed and look out after it. I lay cold and shivering till morning. At breakfast he was sharp and fierce in manner, and made no allusion to the adventure of the night. Neither did I, but I gave an excuse for coming to town—and here I am.'

Holmes looked thoroughly surprised at Miss Presbury's narrative.

'My dear young lady, you say that your room is on the second floor. Is there a long ladder in the garden?'

'No, Mr Holmes; that is the amazing part of it. There is no possible way of reaching the window—and yet he was there.'

'The date being September 5th,' said Holmes. 'That certainly complicates matters.'

It was the young lady's turn to look surprised. 'This is the second time that you have alluded to the date, Mr Holmes,' said Bennett. 'Is it possible that it has any bearing upon the case?'

'It is possible—very possible—and yet I have not my full material at present.'

'Possibly you are thinking of the connection between insanity and phases of the moon?'

'No, I assure you. It was quite a different line of thought. Possibly you can leave your notebook with me and I will check the dates. Now I think, Watson, that our line of action is perfectly clear. This young lady has informed us—and I have the greatest confidence in her intuition—that her father remembers little or nothing which occurs upon certain dates. We will therefore call upon him as if he had given us an appointment upon such a date. He will put it down to his own lack of memory. Thus we will open our campaign by having a good close view of him.'

'That is excellent,' said Mr Bennett. 'I warn you, however, that the Professor is irascible and violent at times.'

Holmes smiled. 'There are reasons why we should come at once—very cogent reasons if my theories hold good. To-morrow, Mr Bennett, will certainly see us in Camford. There is, if I remember right, an inn called the "Chequers" where the port used to be above mediocrity, and the linen was above reproach. I think, Watson, that our lot for the next few days might lie in less pleasant places.'

Monday morning found us on our way to the famous University town—an easy effort on the part of Holmes, who had no roots to pull up, but one which involved frantic planning and hurrying on my part, as my practice was by this time not inconsiderable. Holmes made no allusion to the case until after we had deposited our suit-cases at the ancient hostel of which he had spoken.

'I think, Watson, that we can catch the Professor just before lunch. He lectures at eleven, and should have an interval at home.'

'What possible excuse have we for calling?'

Holmes glanced at his notebook.

'There was a period of excitement upon August 25th.* We will assume that he is a little hazy as to what he does at such times. If we insist that we are there by appointment I think he will hardly venture to contradict us. Have you the effrontery necessary to put it through?'

'We can but try.'

'Excellent, Watson! Compound of the Busy Bee and Excelsior. We can but try—the motto of the firm. A friendly native will surely guide us.'

Such a one on the back of a smart hansom swept us past a row of ancient colleges, and finally turning into a tree-lined drive pulled up at the door of a charming house, girt round with lawns and covered with purple wistaria.* Professor Presbury was certainly surrounded with every sign not only of comfort but of luxury. Even as we pulled up a grizzled head appeared at the front window, and we were aware of a pair of keen eyes from under shaggy brows which surveyed us through large horn glasses. A moment later we were actually in his sanctum, and the mysterious scientist, whose vagaries had brought us from London, was standing before us. There was certainly no sign of eccentricity either in his manner or appearance, for he was a portly, large-featured man, grave, tall, and frock-coated, with the dignity of bearing which a lecturer needs. His eyes were his most remarkable feature, keen, observant, and clever to the verge of cunning.

He looked at our cards. 'Pray sit down, gentlemen. What can I do for you?'

Mr Holmes smiled amiably.

'It was the question which I was about to put to you, Professor.'

'To me, sir!'

'Possibly there is some mistake. I heard through a second person that Professor Presbury of Camford had need of my services.'

'Oh, indeed!' It seemed to me that there was a malicious sparkle in the intense grey eyes. 'You heard that, did you? May I ask the name of your informant?'

'I am sorry, Professor, but the matter was rather confidential. If I have made a mistake there is no harm done. I can only express my regret.'

'Not at all. I should wish to go further into this matter. It interests me. Have you any scrap of writing, any letter or telegram, to bear out your assertion?'

'No, I have not.'

'I presume that you do not go so far as to assert that I summoned you?'

'I would rather answer no questions,' said Holmes.

'No, I dare say not,' said the Professor, with asperity. 'However, that particular one can be answered very easily without your aid.'

He walked across the room to the bell. Our London friend, Mr Bennett, answered the call.

'Come in, Mr Bennett. These two gentlemen have come from London under the impression that they have been summoned. You handle all my correspondence. Have you a note of anything going to a person named Holmes?'

'No, sir,' Bennett answered, with a flush.

'That is conclusive,' said the Professor, glaring angrily at my companion. 'Now, sir'—he leaned forward with his two hands upon the table—'it seems to me that your position is a very questionable one.'

Holmes shrugged his shoulders.

'I can only repeat that I am sorry that we have made a needless intrusion.'

'Hardly enough, Mr Holmes!' the old man cried, in a high screaming voice, with extraordinary malignancy upon his face. He got between us and the door as he spoke, and he shook his two hands at us with furious passion. 'You can hardly get out of it so easily as that.' His face was convulsed and he grinned and gibbered at us in his senseless rage.* I am convinced that we should have had to fight our way out of the room if Mr Bennett had not intervened.

'My dear Professor,' he cried, 'consider your position! Consider the scandal at the University! Mr Holmes is a well-known man. You cannot possibly treat him with such discourtesy.'

Sulkily our host—if I may call him so—cleared the path to the door. We were glad to find ourselves outside the house, and in the quiet of the tree-lined drive. Holmes seemed greatly amused by the episode.

'Our learned friend's nerves are somewhat out of order,' said he. 'Perhaps our intrusion was a little crude, and yet we have gained that personal contact which I desired. But, dear me, Watson, he is surely at our heels. The villain still pursues us.'

There were the sounds of running feet behind, but it was, to my relief, not the formidable Professor but his assistant who appeared round the curve of the drive. He came panting up to us.

'I am so sorry, Mr Holmes. I wished to apologize.'

'My dear sir, there is no need. It is all in the way of professional experience.'

'I have never seen him in a more dangerous mood. But he grows more sinister. You can understand now why his daughter and I are alarmed. And yet his mind is perfectly clear.'

'Too clear!' said Holmes. 'That was my miscalculation. It is evident that his memory is much more reliable than I had thought. By the way, can we, before we go, see the window of Miss Presbury's room?'

Mr Bennett pushed his way through some shrubs and we had a view of the side of the house.

'It is there. The second on the left.'

'Dear me, it seems hardly accessible. And yet you will observe that there is a creeper below and a water-pipe above which give some foothold.'

'I could not climb it myself,' said Mr Bennett.

'Very likely. It would certainly be a dangerous exploit for any normal man.'

'There was one other thing I wished to tell you, Mr Holmes. I have the address of the man in London to whom the Professor writes. He seems to have written this morning and I got it from his blotting-paper. It is an ignoble position for a trusted secretary, but what else can I do?'

Holmes glanced at the paper and put it into his pocket.

'Dorak—a curious name. Slavonic,* I imagine. Well, it is an important link in the chain. We return to London this afternoon, Mr Bennett. I see no good purpose to be served by our remaining. We cannot arrest the Professor, because he has done no crime, nor can we place him under constraint, for he cannot be proved to be mad. No action is as yet possible.'

'Then what on earth are we to do?'

'A little patience, Mr Bennett. Things will soon develop. Unless I am mistaken next Tuesday may mark a crisis. Certainly we shall be in Camford on that day. Meanwhile, the general position is undeniably unpleasant, and if Miss Presbury can prolong her visit—'

'That is easy.'

'Then let her stay till we can assure her that all danger is past. Meanwhile, let him have his way and do not cross him. So long as he is in a good humour all is well.'

'There he is!' said Bennett, in a startled whisper. Looking between the branches we saw the tall, erect figure emerge from the hall door and look around him. He stood leaning forward, his hands swinging straight before him, his head turning from side to side. The secretary with a last wave slipped off among the trees, and we saw him presently rejoin his employer, the two entering the house together in what seemed to be animated and even excited conversation.

'I expect the old gentleman has been putting two and two together,' said Holmes, as we walked hotelwards. 'He struck me as having a particularly clear and logical brain, from the little I saw of him. Explosive, no doubt, but then from his point of view he has something to explode about if detectives are put on his track and he suspects his own household of doing it. I rather fancy that friend Bennett is in for an uncomfortable time.'

Holmes stopped at a post office and sent off a telegram on our way. The answer reached us in the evening, and he tossed it across to me. 'Have visited the Commercial Road* and seen Dorak. Suave person, Bohemian,* elderly. Keeps large general store.—Mercer.'*

'Mercer is since your time,' said Holmes. 'He is my general utility man who looks up routine business. It was important to know something of the man with whom our Professor was so secretly corresponding. His nationality connects up with the Prague visit.'

'Thank goodness that something connects with something,' said I. 'At present we seem to be faced by a long series of inexplicable incidents with no bearing upon each other. For example what possible connection can there be between an angry wolf-hound and a visit to Bohemia, or either of them with a man crawling down a passage at night? As to your dates, that is the biggest mystification of all.'

Holmes smiled and rubbed his hands. We were, I may say, seated in the old sitting-room of the ancient hotel, with a bottle of the famous vintage* of which Holmes had spoken on the table between us.

'Well, now, let us take the dates first,' said he, his finger-tips together and his manner as if he were addressing a class. 'This excellent young man's diary shows that there was trouble upon July 2nd, and from then onwards it seems to have been at nine-day intervals, with, so far as I remember, only one exception. Thus the last outbreak upon Friday was on September 3rd, which also falls into the series, as did August 25th, which preceded it. The thing is beyond coincidence.'

I was forced to agree.

'Let us, then, form the provisional theory that every nine days the Professor takes some strong drug which has a pass- ing but highly poisonous effect. His naturally violent nature is intensified by it. He learned to take this drug while he was in Prague, and is now supplied with it by a Bohemian intermediary in London. This all hangs together, Watson!'

'But the dog, the face at the window, the creeping man in the passage?'

'Well, well, we have made a beginning. I should not expect any fresh developments until next Tuesday. In the meantime we can only keep in touch with friend Bennett and enjoy the amenities of this charming town.'*

In the morning Mr Bennett slipped round to bring us the latest report. As Holmes had imagined, times had not been easy with him. Without exactly accusing him of being responsible for our presence, the Professor had been very rough and rude in his speech, and evidently felt some strong grievance. This morning he was quite himself again, however, and had delivered his usual brilliant lecture to a crowded class. 'Apart from his queer fits,' said Bennett, 'he has actually more energy and vitality than I can ever remember, nor was his brain ever clearer. But it's not he—it's never the man whom we have known.'

'I don't think you have anything to fear now for a week at least,' Holmes answered. 'I am a busy man, and Dr Watson has his patients to attend to. Let us agree that we meet here at this hour next Tuesday, and I shall be surprised if before we leave you again we are not able to explain, even if we cannot perhaps put an end to, your troubles. Meanwhile, keep us posted in what occurs.'

I saw nothing of my friend for the next few days, but on the following Monday evening I had a short note asking me to meet him next day at the train. From what he told me as we travelled up to Camford all was well, the peace of the Professor's house had been unruffled, and his own conduct perfectly normal. This also was the report which was given us by Mr Bennett himself when he called upon us that evening at our old quarters in the 'Chequers'. 'He heard from his London correspondent today. There was a letter and there was a small packet, each with the cross under the stamp which warned me not to touch them. There has been nothing else.'

'That may prove quite enough,' said Holmes, grimly. 'Now, Mr Bennett, we shall, I think, come to some conclusion to-night. If my deductions are correct we should have an opportunity of bringing matters to a head. In order to do so it is necessary to hold the Professor under observation. I would suggest, therefore, that you remain awake and on the look-out. Should you hear him pass your door do not interrupt him, but follow him as discreetly as you can. Dr Watson and I will not be far off. By the way, where is the key of that little box of which you spoke?'

'Upon his watch-chain.'

'I fancy our researches must lie in that direction. At the worst the lock should not be very formidable. Have you any other able-bodied man on the premises?'

'There is the coachman, Macphail.'

'Where does he sleep?'

'Over the stables.'

'We might possibly want him. Well, we can do no more until we see how things develop. Good-bye—but I expect that we shall see you before morning.'

It was nearly midnight before we took our station among some bushes immediately opposite the hall door of the Professor. It was a fine night, but chilly, and we were glad of our warm overcoats. There was a breeze, and clouds were scudding across the sky, obscuring from time to time the half-moon. It would have been a dismal vigil were it not for the expectation and excitement which carried us along, and the assurance of my comrade that we had probably reached the end of the strange sequence of events which had engaged our attention.

'If the cycle of nine days holds good then we shall have the Professor at his worst tonight,' said Holmes. 'The fact that these strange symptoms began after his visit to Prague, that he is in secret correspondence with a Bohemian dealer in London, who presumably represents someone in Prague, and that he received a packet from him this very day, all point in one direction. What he takes and why he takes it are still beyond our ken, but that it emanates in some way from Prague is clear enough. He takes it under definite directions which regulate this ninth-day system, which was the first point which attracted my attention. But his symptoms are most remarkable. Did you observe his knuckles?'

I had to confess that I did not.

'Thick and horny in a way which is quite new in my experience. Always look at the hands first, Watson. Then cuffs, trouser-knees, and boots. Very curious knuckles which can only be explained by the mode of progression observed by—' Holmes paused, and suddenly clapped his hand to his forehead. 'Oh, Watson, Watson, what a fool I have been! It seems incredible, and yet it must be true. All points in one direction. How could I miss seeing the connection of ideas? Those knuckles—how could I have passed those knuckles? And the dog! And the ivy! It's surely time that I disappeared into that little farm of my dreams. Look out, Watson! Here he is! We shall have the chance of seeing for ourselves.'

The hall door had slowly opened, and against the lamp-lit background we saw the tall figure of Professor Presbury. He was clad in his dressing-gown. As he stood outlined in the doorway he was erect but leaning forward with dangling arms, as when we saw him last.

Now he stepped forward into the drive, and an extraordinary change came over him. He sank down into a crouching position, and moved along upon his hands and feet, skipping every now and then as if he were overflowing with energy and vitality. He moved along the face of the house and then round the corner. As he disappeared Bennett slipped through the hall door and softly followed him.

'Come, Watson, come!' cried Holmes, and we stole as softly as we could through the bushes until we had gained a spot whence we could see the other side of the house, which was bathed in the light of the half-moon. The Professor was clearly visible crouching at the foot of the ivy-covered wall. As we watched him he suddenly began with incredible agility to ascend it. From branch to branch he sprang, sure of foot and firm of grasp, climbing apparently in mere joy at his own powers, with no definite object in view. With his dressing-gown flapping on each side of him he looked like some huge bat glued against the side of his own house, a great square dark patch upon the moonlit wall. Presently he tired of this amusement, and dropping from branch to branch, he squatted down into the old attitude and moved towards the stables, creeping along in the same strange way as before. The wolf-hound was out now, barking furiously, and more excited than ever when it actually caught sight of its master. It was straining on its chain, and quivering with eagerness and rage. The Professor squatted down very deliberately just out of reach of the hound, and began to provoke it in every possible way. He took handfuls of pebbles from the drive and threw them in the dog's face, prodded him with a stick which he had picked up, flicked his hands about only a few inches from the gaping mouth, and endeavoured in every way to increase the animal's fury, which was already beyond all control. In all our adventures I do not know that I have ever seen a more strange sight than this impassive and still dignified figure crouching froglike upon the ground and goading to a wilder exhibition of passion the maddened hound, which ramped* and raged in front of him, by all manner of ingenious and calculated cruelty.

And then in a moment it happened! It was not the chain that broke, but it was the collar that slipped, for it had been made for a thick-necked Newfoundland.* We heard the rattle of falling metal, and the next instant dog and man were rolling on the ground together, the one roaring in rage, the other screaming in a strange shrill falsetto of terror. It was a very narrow thing for the Professor's life. The savage creature had him fairly by the throat, its fangs had bitten deep, and he was senseless before we could reach them and drag the two apart. It might have been a dangerous task for us, but Bennett's voice and presence brought the great wolf-hound instantly to reason. The uproar had brought the sleepy and astonished coachman from his room above the stables. 'I'm not surprised,' said he, shaking his head. 'I've seen him at it before. I knew the dog would get him sooner or later.'

The hound was secured, and together we carried the Professor up to his room, where Bennett, who had a medical degree, helped me to dress his torn throat. The sharp teeth had passed dangerously near the carotid artery, and the haemorrhage was serious. In half an hour the danger was past, I had given the patient an injection of morphia, and he had sunk into deep sleep. Then, and only then, were we able to look at each other and to take stock of the situation.

'I think a first-class surgeon should see him,' said I.

'For God's sake, no!' cried Bennett. 'At present the scandal is confined to our own household. It is safe with us. If it gets beyond these walls it will never stop. Consider his position at the University, his European reputation, the feelings of his daughter.'

'Quite so,' said Holmes. 'I think it may be quite possible to keep the matter to ourselves, and also to prevent its recurrence now that we have a free hand. The key from the watch-chain, Mr Bennett. Macphail will guard the patient and let us know if there is any change. Let us see what we can find in the Professor's mysterious box.'

There was not much, but there was enough—an empty phial,* another nearly full, a hypodermic syringe, several letters in a crabbed, foreign hand. The marks on the envelopes showed that they were those which had disturbed the routine of the secretary, and each was dated from the Commercial Road and signed 'A. Dorak'. They were mere invoices to say that a fresh bottle was being sent to Professor Presbury, or receipts to acknowledge money. There was one other envelope, however, in a more educated hand and bearing the Austrian stamp* with the postmark of Prague. 'Here we have our material!' cried Holmes, as he tore out the enclosure.

HONOURED COLLEAGUE, [it ran]

Since your esteemed visit I have thought much of your case, and though in your circumstances there are some special reasons for the treatment, I would none the less enjoin caution, as my results have shown that it is not without danger of a kind.

It is possible that the serum* of Anthropoid* would have been better. I have, as I explained to you, used black-faced Langur* because a specimen was accessible. Langur is, of course, a crawler and climber, while Anthropoid walks erect, and is in all ways nearer.

I beg you to take every possible precaution that there be no premature revelation of the process. I have one other client in England, and Dorak is my agent for both.

Weekly reports will oblige.

Yours with high esteem,

H. LOWENSTEIN

Lowenstein! The name brought back to me the memory of some snippet from a newspaper which spoke of an obscure scientist who was striving in some unknown way for the secret of rejuvenescence and the elixir of life.* Lowenstein of Prague! Lowenstein with the wondrous strength-giving serum, tabooed by the profession because he refused to reveal its source. In a few words I said what I remembered. Bennett had taken a manual of zoology from the shelves. ' "Langur",' he read, ' "the great black-faced monkey of the Himalayan slopes, biggest and most human of climbing monkeys." Many details are added. Well, thanks to you, Mr Holmes, it is very clear that we have traced the evil to its source.'

'The real source,' said Holmes, 'lies, of course, in that untimely love affair which gave our impetuous Professor the idea that he could only gain his wish by turning himself into a younger man. When one tries to rise above Nature one is liable to fall below it. The highest type of man may revert to the animal if he leaves the straight road of destiny.' He sat musing for a little with the phial in his hand, looking at the clear liquid within. 'When I have written to this man and told him that I hold him criminally responsible for the poisons which he circulates, we will have no more trouble. But it may recur. Others may find a better way. There is danger there—a very real danger to humanity. Consider, Watson, that the material, the sensual, the worldly would all prolong their worthless lives. The spiritual would not avoid the call to something higher. It would be the survival of the least fit. What sort of cesspool may not our poor world become?'* Suddenly the dreamer disappeared, and Holmes, the man of action, sprang from his chair. 'I think there is nothing more to be said, Mr Bennett. The various incidents will now fit themselves easily into the general scheme. The dog, of course, was aware of the change far more quickly than you. His smell* would ensure that. It was the monkey, not the Professor, whom Roy attacked, just as it was the monkey who teased Roy. Climbing was a joy to the creature, and it was a mere chance, I take it, that the pastime brought him to the young lady's window. There is an early train to town, Watson, but I think we shall just have time for a cup of tea at the "Chequers" before we catch it.'

The Sussex Vampire

HOLMES had read carefully a note which the last post had brought him. Then, with the dry chuckle which was his nearest approach to a laugh, he tossed it over to me.

'For a mixture of the modern and the mediaeval, of the practical and of the wildly fanciful, I think this is surely the limit,' said he. 'What do you make of it, Watson?'

I read as follows:

46 OLD JEWRY*
Nov. 19th.

Re Vampires

SIR,

Our client, Mr Robert Ferguson, of Ferguson & Muirhead, tea brokers, of Mincing Lane,* has made some inquiry from us in a communication of even date concerning vampires. As our firm specializes entirely upon the assessment of machinery the matter hardly comes within our purview, and we have therefore recommended Mr Ferguson to call upon you and lay the matter before you. We have not forgotten your successful action in the case of Matilda Briggs.

We are, Sir,
Faithfully yours,
MORRISON, MORRISON, AND DODD
per E.J.C.*

'Matilda Briggs was not the name of a young woman, Watson,' said Holmes, in a reminiscent voice. 'It was a ship which is associated with the giant rat of Sumatra,* a story for which the world is not yet prepared. But what do we know about vampires? Does it come within our purview either? Anything is better than stagnation, but really we seem to have been switched on to a Grimm's fairy tale.* Make a long arm, Watson, and see what V has to say.'

I leaned back and took down the great index volume to which he referred. Holmes balanced it on his knee and his eyes moved slowly and lovingly over the record of old cases, mixed with the accumulated information of a lifetime.

'Voyage of the Gloria Scott,'* he read. 'That was a bad business. I have some recollection that you made a record of it, Watson, though I was unable to congratulate you upon the result. Victor Lynch,* forger. Venomous lizard or gila. Remarkable case, that! Vittoria, the circus belle. Vanderbilt and the Yeggman.* Vipers. Vigor,* the Hammersmith wonder. Hullo! Hullo! Good old index. You can't beat it. Listen to this, Watson. Vampirism* in Hungary. And again, Vampires in Transylvania,'* He turned over the pages with eagerness, but after a short intent perusal he threw down the great book with a snarl of disappointment.

'Rubbish, Watson, rubbish! What have we to do with walking corpses who can only be held in their grave by stakes driven through their hearts? It's pure lunacy.'*

'But surely,' said I, 'the vampire was not necessarily a dead man? A living person might have the habit. I have read for example, of the old sucking the blood of the young in order to retain their youth.'

'You are right, Watson. It mentions the legend in one of these references. But are we to give serious attention to such things? This agency* stands flat-footed upon the ground, and there it must remain. The world is big enough for us. No ghosts need apply. I fear that we cannot take Mr Robert Ferguson very seriously. Possibly this note may be from him, and may throw some light upon what is worrying him.'

He took up a second letter which had lain unnoticed upon the table whilst he had been absorbed with the first. This he began to read with a smile of amusement upon his face which gradually faded away into an expression of intense interest and concentration. When he had finished he sat for some little time lost in thought with the letter dangling from his fingers. Finally, with a start, he aroused himself from his reverie.

'Cheeseman's, Lamberley. Where is Lamberley, Watson?'

'It is in Sussex, south of Horsham.'*

'Not very far, eh? And Cheeseman's?'

'I know that country, Holmes. It is full of old houses which are named after the men who built them centuries ago. You get Odley's and Harvey's and Carriton's—the folk are forgotten but their names live in their houses.'

'Precisely,' said Holmes coldly. It was one of the peculiarities of his proud, self-contained nature that, though he docketed any fresh information very quickly and accurately in his brain, he seldom made any acknowledgement to the giver. 'I rather fancy we shall know a good deal more about Cheeseman's, Lamberley, before we are through. The letter is, as I had hoped, from Robert Ferguson. By the way, he claims acquaintance with you.'

'With me!'

'You had better read it.'

He handed the letter across. It was headed with the address quoted.

DEAR MR HOLMES, [it said]

I have been recommended to you by my lawyers, but indeed the matter is so extraordinarily delicate that it is most difficult to discuss. It concerns a friend for whom I am acting. This gentleman married some five years ago a Peruvian lady, the daughter of a Peruvian merchant, whom he had met in connection with the importation of nitrates. The lady was very beautiful, but the fact of her foreign birth and of her alien religion* always caused a separation of interests and of feelings between husband and wife, so that after a time his love may have cooled towards her and he may have come to regard their union as a mistake. He felt there were sides of her character which he could never explore or understand. This was the more painful as she was as loving a wife as a man could have—to all appearance absolutely devoted.

Now for the point which I will make more plain when we meet. Indeed, this note is merely to give you a general idea of the situation and to ascertain whether you would care to interest yourself in the matter. The lady began to show some curious traits, quite alien to her ordinarily sweet and gentle disposition. The gentleman had been married twice and he had one son by the first wife. This boy was now fifteen, a very charming and affectionate youth, though unhappily injured through an accident in childhood. Twice the wife was caught in the act of assaulting this poor lad in the most unprovoked way. Once she struck him with a stick and left a great weal on his arm.

This was a small matter, however, compared with her conduct to her own child, a dear boy just under one year of age. On one occasion about a month ago this child had been left by its nurse for a few minutes. A loud cry from the baby, as of pain, called the nurse back. As she ran into the room she saw her employer, the lady, leaning over the baby and apparently biting his neck. There was a small wound in the neck, from which a stream of blood had escaped. The nurse was so horrified that she wished to call the husband, but the lady implored her not to do so, and actually gave her five pounds as a price for her silence. No explanation was ever given, and for the moment the matter was passed over.

It left, however, a terrible impression upon the nurse's mind, and from that time she began to watch her mistress closely, and to keep a closer guard upon the baby, whom she tenderly loved. It seemed to her that even as she watched the mother, so the mother watched her, and that every time she was compelled to leave the baby alone the mother was waiting to get at it. Day and night the nurse covered the child, and day and night the silent, watchful mother seemed to be lying in wait as a wolf waits for a lamb. It must read most incredible to you, and yet I beg you to take it seriously, for a child's life and a man's sanity may depend upon it.

At last there came one dreadful day when the facts could no longer be concealed from the husband. The nurse's nerve had given way; she could stand the strain no longer, and she made a clean breast of it all to the man. To him it seemed as wild a tale as it may now seem to you. He knew his wife to be a loving wife, and, save for the assaults upon her stepson, a loving mother. Why, then, should she wound her own dear little baby? He told the nurse that she was dreaming, that her suspicions were those of a lunatic, and that such libels upon her mistress were not to be tolerated. Whilst they were talking, a sudden cry of pain was heard. Nurse and master rushed together to the nursery. Imagine his feelings, Mr Holmes, as he saw his wife rise from a kneeling position beside the cot, and saw blood upon the child's exposed neck and upon the sheet. With a cry of horror, he turned his wife's face to the light and saw blood all round her lips. It was she—she beyond all question—who had drunk the poor baby's blood.

So the matter stands. She is now confined to her room. There has been no explanation. The husband is half demented. He knows, and I know, little of Vampirism beyond the name. We had thought it was some wild tale of foreign parts. And yet here in the very heart of the English Sussex—well, all this can be discussed with you in the morning. Will you see me? Will you use your great powers in aiding a distracted man? If so, kindly wire to Ferguson, Cheeseman's, Lamberley, and I will be at your rooms by ten o'clock.

Yours faithfully,
ROBERT FERGUSON

PS.—I believe your friend Watson played Rugby for Blackheath* when I was three-quarter for Richmond.* It is the only personal introduction which I can give.

'Of course I remember him,' said I, as I laid down the letter. 'Big Bob Ferguson, the finest three-quarter Richmond ever had. He was always a good-natured chap. It's like him to be so concerned over a friend's case.'

Holmes looked at me thoughtfully and shook his head.

'I never get your limits, Watson,' said he. 'There are unexplored possibilities about you. Take a wire* down, like a good fellow. "Will examine your case with pleasure." '

'*Your* case!'

'We must not let him think that this agency is a home for the weak-minded. Of course it is his case. Send him that wire and let the matter rest till morning.'

Promptly at ten o'clock next morning Ferguson strode into our room. I had remembered him as a long, slab-sided man with loose limbs and a fine turn of speed, which had carried him round many an opposing back. There is surely nothing in life more painful than to meet the wreck of a fine athlete whom one has known in his prime. His great frame had fallen in, his flaxen hair was scanty, and his shoulders were bowed. I fear that I roused corresponding emotions in him.

'Hullo, Watson,' said he, and his voice was still deep and hearty. 'You don't look quite the man you did when I threw you over the ropes into the crowd at the Old Deer Park.* I expect I have changed a bit also. But it's this last day or two that has aged me. I see by your telegram, Mr Holmes, that it is no use my pretending to be anyone's deputy.'

'It is simpler to deal direct,' said Holmes.

'Of course it is. But you can imagine how difficult it is when you are speaking of the one woman you are bound to protect and help. What can I do? How am I to go to the police with such a story? And yet the kiddies have got to be protected. Is it madness, Mr Holmes? Is it something in the blood? Have you any similar case in your experience? For God's sake, give me some advice, for I am at my wits' end.'

'Very naturally, Mr Ferguson. Now sit here and pull yourself together and give me a few clear answers. I can assure you that I am very far from being at my wits' end, and that I am confident we shall find some solution. First of all, tell me what steps you have taken. Is your wife still near the children?'

'We had a dreadful scene. She is a most loving woman, Mr Holmes. If ever a woman loved a man with all her heart and soul, she loves me. She was cut to the heart that I should have discovered this horrible, this incredible, secret. She would not even speak. She gave no answer to my reproaches, save to gaze at me with a sort of wild, despairing look in her eyes. Then she rushed to her room and locked herself in. Since then she has refused to see me. She has a maid who was with her before her marriage, Dolores by name—a friend rather than a servant. She takes her food to her.'

'Then the child is in no immediate danger?'

'Mrs Mason, the nurse, has sworn that she will not leave it night or day. I can absolutely trust her. I am more uneasy about poor little Jack, for, as I told you in my note, he has twice been assaulted by her.'

'But never wounded?'

'No; she struck him savagely. It is the more terrible as he is a poor little inoffensive cripple.' Ferguson's gaunt features softened as he spoke of his boy. 'You would think that the dear lad's condition would soften anyone's heart. A fall in childhood and a twisted spine, Mr Holmes. But the dearest, most loving heart within.'

Holmes had picked up the letter of yesterday and was reading it over. 'What other inmates are there in your house, Mr Ferguson?'

'Two servants who have not been long with us. One stable-hand, Michael, who sleeps in the house. My wife, myself, my boy Jack, baby, Dolores, and Mrs Mason. That is all.'

'I gather that you did not know your wife well at the time of your marriage?'

'I had only known her a few weeks.'

'How long had this maid Dolores been with her?'

'Some years.'

'Then your wife's character would really be better known by Dolores than by you?'

'Yes, you may say so.'

Holmes made a note.

'I fancy', said he, 'that I may be of more use at Lamberley than here. It is eminently a case for personal investigation. If the lady remains in her room, our presence could not annoy or inconvenience her. Of course, we would stay at the inn.'

Ferguson gave a gesture of relief.

'It is what I hoped, Mr Holmes. There is an excellent train at two from Victoria, if you could come.'

'Of course we could come. There is a lull at present. I can give you my undivided energies. Watson, of course, comes with us. But there are one or two points upon which I wish to be very sure before I start. This unhappy lady as I understand it, has appeared to assault both the children, her own baby and your little son?'

'That is so.'

'But the results take different forms, do they not? She has beaten your son.'

'Once with a stick and once very savagely with her hands.'

'Did she give no explanation why she struck him?'

'None, save that she hated him. Again and again she said so.'

'Well, that is not unknown among stepmothers. A posthumous jealousy, we will say. Is the lady jealous by nature?'

'Yes, she is very jealous—jealous with all the strength of her fiery tropical love.'

'But the boy—he is fifteen, I understand, and probably very developed in mind, since his body has been circumscribed in action. Did he give you no explanation of these assaults?'

'No; he declared there was no reason.'

'Were they good friends at other times?'

'No; there was never any love between them.'

'Yet you say he is affectionate?'

'Never in the world could there be so devoted a son. My life is his life. He is absorbed in what I say or do.'

Once again Holmes made a note. For some time he sat lost in thought.

'No doubt you and the boy were great comrades before this second marriage. You were thrown very close together, were you not?'

'Very much so.'

'And the boy, having so affectionate a nature, was devoted, no doubt, to the memory of his mother?'

'Most devoted.'

'He would certainly seem to be a most interesting lad. There is one other point about these assaults. Were the strange attacks upon the baby and the assaults upon your son at the same period?'

'In the first case it was so. It was as if some frenzy had seized her, and she had vented her rage upon both. In the second case it was only Jack who suffered. Mrs Mason had no complaint to make about the baby.'

'That certainly complicates matters.'

'I don't quite follow you, Mr Holmes.'

'Possibly not. One forms provisional theories and waits for time or fuller knowledge to explode them. A bad habit, Mr Ferguson; but human nature is weak. I fear that your old friend here has given an exaggerated view of my scientific methods. However, I will only say at the present stage that your problem does not appear to me to be insoluble, and that you may expect to find us at Victoria at two o'clock.'

It was evening of a dull, foggy November day when, having left our bags at the 'Chequers', Lamberley, we drove through the Sussex clay of a long winding lane, and finally reached the isolated and ancient farmhouse in which Ferguson dwelt. It was a large, straggling building, very old in the centre, very new at the wings, with towering Tudor chimneys* and a lichen-spotted,* high-pitched roof of Horsham slabs. The doorsteps were worn into curves, and the ancient tiles which lined the porch were marked with the rebus* of a cheese and a man, after the original builder. Within, the ceilings were corrugated with heavy oaken beams, and the uneven floors sagged into sharp curves. An odour of age and decay pervaded the whole crumbling building.

There was one very large central room, into which Ferguson led us. Here, in a huge old-fashioned fireplace with an iron screen behind it dated 1670, there blazed and spluttered a splendid log fire.

The room, as I gazed round, was a most singular mixture of dates and of places. The half-panelled* walls may well have belonged to the original yeoman farmer* of the seventeenth century. They were ornamented, however, on the lower part by a line of well-chosen modern water-colours; while above, where yellow plaster took the place of oak, there was hung a fine collection of South American utensils and weapons, which had been brought, no doubt, by the Peruvian lady upstairs. Holmes rose, with that quick curiosity which sprang from his eager mind, and examined them with some care. He returned with his eyes full of thought.

'Hullo!' he cried. 'Hullo!'

A spaniel had lain in a basket in the corner. It came slowly forward towards its master, walking with difficulty. Its hind legs moved irregularly and its tail was on the ground.* It licked Ferguson's hand.

'What is it, Mr Holmes?'

'The dog. What's the matter with it?'

'That's what puzzled the vet. A sort of paralysis. Spinal meningitis, he thought. But it is passing. He'll be all right soon—won't you, Carlo?'

A shiver of assent passed through the drooping tail. The dog's mournful eyes passed from one of us to the other. He knew that we were discussing his case.

'Did it come on suddenly?'

'In a single night.'

'How long ago?'

'It may have been four months ago.'

'Very remarkable. Very suggestive.'

'What do you see in it, Mr Holmes?'

'A confirmation of what I had already thought.'

'For God's sake, what *do* you think, Mr Holmes? It may be a mere intellectual puzzle to you, but it is life and death to me! My wife a would-be murderer—my child in constant danger! Don't play with me, Mr Holmes. It is too terribly serious.'

The big Rugby three-quarter was trembling all over. Holmes put his hand soothingly upon his arm.

'I fear that there is pain for you, Mr Ferguson, whatever the solution may be,' said he. 'I would spare you all I can. I cannot say more for the instant, but before I leave this house I hope I may have something definite.'

'Please God you may! If you will excuse me, gentlemen, I will go up to my wife's room and see if there has been any change.'

He was away some minutes, during which Holmes resumed his examination of the curiosities upon the wall. When our host returned it was clear from his downcast face that he had made no progress. He brought with him, a tall, slim, brown-faced girl.

'The tea is ready, Dolores,' said Ferguson. 'See that your mistress has everything she can wish.'

'She verra ill,' cried the girl, looking with indignant eyes at her master. 'She no ask for food. She verra ill. She need doctor. I frightened stay alone with her without doctor.'

Ferguson looked at me with a question in his eyes.

'I should be so glad if I could be of use.'

'Would your mistress see Dr Watson?'

'I take him. I no ask leave. She needs doctor.'

'Then I'll come with you at once.'

I followed the girl, who was quivering with strong emotion, up the staircase and down an ancient corridor. At the end was an iron-clamped and massive door. It struck me as I looked at it that if Ferguson tried to force his way to his wife he would find it no easy matter. The girl drew a key from her pocket, and the heavy oaken planks creaked upon their old hinges. I passed in and she swiftly followed, fastening the door behind her.

On the bed a woman was lying who was clearly in a high fever. She was only half conscious, but as I entered she raised a pair of frightened but beautiful eyes and glared at me in apprehension. Seeing a stranger, she appeared to be relieved, and sank back with a sigh upon the pillow. I stepped up to her with a few reassuring words, and she lay still while I took her pulse and temperature. Both were high, and yet my impression was that the condition was rather that of mental and nervous excitement than of any actual seizure.

'She lie like that one day, two day. I 'fraid she die,' said the girl.

The woman turned her flushed and handsome face towards me.

'Where is my husband?'

'He is below, and would wish to see you.'

'I will not see him. I will not see him.' Then she seemed to wander off into delirium. 'A fiend! A fiend! Oh, what shall I do with this devil?'

'Can I help you in any way?'

'No. No one can help. It is finished. All is destroyed. Do what I will, all is destroyed.'

The woman must have some strange delusion. I could not see honest Bob Ferguson in the character of fiend or devil.

'Madame,' I said, 'your husband loves you dearly. He is deeply grieved at this happening.'

Again she turned on me those glorious eyes.

'He loves me. Yes. But do I not love him? Do I not love him even to sacrifice myself rather than break his dear heart. That is how I love him. And yet he could think of me—he could speak to me so.'

'He is full of grief, but he cannot understand.'

'No, he cannot understand. But he should trust.'

'Will you not see him?' I suggested.

'No, no; I cannot forget those terrible words nor the look upon his face. I will not see him. Go now. You can do nothing for me. Tell him only one thing. I want my child. I have a right to my child. That is the only message I can send him.' She turned her face to the wall and would say no more.

I returned to the room downstairs, where Ferguson and Holmes still sat by the fire. Ferguson listened moodily to my account of the interview.

'How can I send her the child?' he said. 'How do I know what strange impulse might come upon her? How can I ever forget how she rose from beside it with its blood on her lips?' He shuddered at the recollection. 'The child is safe with Mrs Mason, and there he must remain.'

A smart maid, the only modern thing which we had seen in the house, had brought in some tea. As she was serving it the door opened and a youth entered the room. He was a remarkable lad, pale-faced and fair-haired, with excitable light blue eyes which blazed into a sudden flame of emotion and joy as they rested upon his father. He rushed forward and threw his arms round his neck with the abandon of a loving girl.

'Oh, daddy,' he cried, 'I did not know that you were due yet. I should have been here to meet you. Oh, I am so glad to see you!'

Ferguson gently disengaged himself from the embrace with some little show of embarrassment.

'Dear old chap,' said he, patting the flaxen head with a very tender hand. 'I came early because my friends, Mr Holmes and Dr Watson, have been persuaded to come down and spend an evening with us.'

'Is that Mr Holmes, the detective?'

'Yes.'

The youth looked at us with a very penetrating and, as it seemed to me, unfriendly gaze.

'What about your other child, Mr Ferguson?' asked Holmes. 'Might we make the acquaintance of the baby?'

'Ask Mrs Mason to bring baby down,' said Ferguson. The boy went off with a curious, shambling gait which told my surgical eyes that he was suffering from a weak spine.* Presently he returned, and behind him came a tall, gaunt woman bearing in her arms a very beautiful child, dark-eyed, golden-haired, a wonderful mixture of the Saxon and the Latin. Ferguson was evidently devoted to it, for he took it into his arms and fondled it most tenderly.

'Fancy anyone having the heart to hurt him,' he muttered, as he glanced down at the small, angry red pucker upon the cherub throat.

It was at this moment that I chanced to glance at Holmes, and saw a most singular intentness in his expression. His face was as set as if it had been carved out of old ivory, and his eyes, which had glanced for a moment at father and child, were now fixed with eager curiosity upon something at the other side of the room. Following his gaze I could only guess that he was looking out through the window at the melancholy, dripping garden. It is true that a shutter had half closed outside and obstructed the view, but none the less it was certainly at the window that Holmes was fixing his concentrated attention. Then he smiled, and his eyes came back to the baby. On its chubby neck there was this small puckered mark. Without speaking, Holmes examined it with care. Finally he shook one of the dimpled fists which waved in front of him.

'Good-bye, little man. You have made a strange start in life. Nurse, I should wish to have a word with you in private.'

He took her aside and spoke earnestly for a few minutes. I only heard the last words, which were: 'Your anxiety will soon, I hope, be set at rest.' The woman, who seemed to be a sour, silent kind of creature, withdrew with the child.

'What is Mrs Mason like?' asked Holmes.

'Not very prepossessing externally, as you can see, but a heart of gold, and devoted to the child.'

'Do you like her, Jack?' Holmes turned suddenly upon the boy. His expressive mobile face shadowed over, and he shook his head.

'Jacky has very strong likes and dislikes,' said Ferguson, putting his arm round the boy. 'Luckily I am one of his likes.'

The boy cooed and nestled his head upon his father's breast. Ferguson gently disengaged him.

'Run away, little Jacky,' said he, and he watched his son with loving eyes until he disappeared. 'Now, Mr Holmes,' he continued, when the boy was gone. 'I really feel that I have brought you on a fool's errand, for what can you possibly do, save give your sympathy? It must be an exceedingly delicate and complex affair from your point of view.'

'It is certainly delicate,' said my friend, with an amused smile, 'but I have not been struck up to now with its complexity. It has been a case for intellectual deduction, but when this original intellectual deduction is confirmed point by point by quite a number of independent incidents, then the subjective becomes objective and we can say confidently that we have reached our goal. I had, in fact, reached it before we left Baker Street, and the rest has merely been observation and confirmation.'

Ferguson put his big hand to his furrowed forehead.*

'For heaven's sake, Holmes,' he said hoarsely, 'if you can see the truth in this matter, do not keep me in suspense. How do I stand? What shall I do? I care nothing as to how you have found your facts so long as you have really got them.'

'Certainly I owe you an explanation, and you shall have it. But you will permit me to handle the matter in my own way? Is the lady capable of seeing us, Watson?'

'She is ill, but she is quite rational.'

'Very good. It is only in her presence that we can clear the matter up. Let us go up to her.'

'She will not see me,' cried Ferguson.

'Oh, yes, she will,' said Holmes. He scribbled a few lines upon a sheet of paper. 'You at least have the *entrée*, Watson. Will you have the goodness to give the lady this note?'

I ascended again and handed the note to Dolores, who cautiously opened the door. A minute later I heard a cry from within, a cry in which joy and surprise seemed to be blended. Dolores looked out.

'She will see them. She will leesten,' said she.

At my summons Ferguson and Holmes came up. As we entered the room Ferguson took a step or two towards his wife, who had raised herself in the bed, but she held out her hand to repulse him. He sank into an arm-chair, while Holmes seated himself beside him, after bowing to the lady, who looked at him with wide-eyed amazement.

'I think we can dispense with Dolores,' said Holmes. 'Oh, very well, madame, if you would rather she stayed I can see no objection. Now, Mr Ferguson, I am a busy man with many calls, and my methods have to be short and direct. The swiftest surgery is the least painful. Let me first say what will ease your mind. Your wife is a very good, a very loving, and a very ill-used woman.'

Ferguson sat up with a cry of joy.

'Prove that, Mr Holmes, and I am your debtor for ever.'

'I will do so, but in doing so I must wound you deeply in another direction.'

'I care nothing so long as you clear my wife. Everything on earth is insignificant compared to that.'

'Let me tell you, then, the train of reasoning which passed through my mind in Baker Street. The idea of a vampire was to me absurd. Such things do not happen in criminal practice in England. And yet your observation was precise. You had seen the lady rise from beside the child's cot with the blood upon her lips.'

'I did.'

'Did it not occur to you that a bleeding wound may be sucked for some other purpose than to draw the blood from it? Was there not a Queen in English history* who sucked such a wound to draw poison from it?'

'Poison!'

'A South American household. My instinct felt the presence of those weapons upon the wall before my eyes ever saw them. It might have been other poison, but that was what occurred to me. When I saw that little empty quiver beside the small bird-bow,* it was just what I expected to see. If the child were pricked with one of those arrows dipped in curare* or some other devilish drug, it would mean death if the venom were not sucked out.

'And the dog! If one were to use such a poison, would one not try it first in order to see that it had not lost its power? I did not foresee the dog, but at least I understood him and he fitted into my reconstruction.

'Now do you understand? Your wife feared such an attack. She saw it made and saved the child's life, and yet she shrank from telling you all the truth, for she knew how you loved the boy and feared lest it break your heart.'

'Jacky!'

'I watched him as you fondled the child just now. His face was clearly reflected in the glass of the window where the shutter formed a background. I saw such jealousy, such cruel hatred, as I have seldom seen in a human face.'

'My Jacky!'

'You have to face it, Mr Ferguson. It is the more painful because it is a distorted love, a maniacal exaggerated love for you, and possibly for his dead mother, which has prompted his action. His very soul is consumed with hatred for this splendid child, whose health and beauty are a contrast to his own weakness.'

'Good God! It is incredible!'

'Have I spoken the truth, madame?'

The lady was sobbing, with her face buried in the pillows. Now she turned to her husband.

'How could I tell you, Bob? I felt the blow it would be to you. It was better that I should wait and that it should come from some other lips than mine. When this gentleman, who seems to have powers of magic, wrote that he knew all, I was glad.'

'I think a year at sea* would be my prescription for Master Jacky,' said Holmes, rising from his chair. 'Only one thing is still clouded, madame. We can quite understand your attacks upon Master Jacky. There is a limit to a mother's patience. But how did you dare to leave the child these last two days?'

'I had told Mrs Mason. She knew.'

'Exactly. So I imagined.'

Ferguson was standing by the bed, choking, his hands outstretched and quivering.

'This, I fancy, is the time for our exit, Watson,' said Holmes in a whisper. 'If you will take one elbow of the too faithful Dolores, I will take the other. There, now,' he added, as he closed the door behind him, 'I think we may leave them to settle the rest among themselves.'

I have only one further note in this case. It is the letter which Holmes wrote in final answer to that with which the narrative begins. It ran thus:

BAKER STREET,
Nov. 21st.

Re Vampires

SIR,

Referring to your letter of the 19th, I beg to state that I have looked into the inquiry of your client, Mr Robert Ferguson, of Ferguson and Muirhead, tea brokers, of Mincing Lane, and that the matter has been brought to a satisfactory conclusion. With thanks for your recommendation,

I am, Sir,
Faithfully yours,
SHERLOCK HOLMES

The Three Garridebs

IT may have been a comedy, or it may have been a tragedy. It cost one man his reason, it cost me a blood-letting, and it cost yet another man the penalties of the law. Yet there was certainly an element of comedy. Well, you shall judge for yourselves.

I remember the date very well, for it was in the same month that Holmes refused a knighthood* for services which may perhaps some day be described. I only refer to the matter in passing, for in my position of partner and confidant I am obliged to be particularly careful to avoid any indiscretion. I repeat, however, that this enables me to fix the date, which was the latter end of June, 1902, shortly after the conclusion of the South African War. Holmes had spent several days in bed, as was his habit from time to time, but he emerged that morning with a long foolscap document in his hand and a twinkle of amusement in his austere grey eyes.

'There is a chance for you to make some money, friend Watson,' said he. 'Have you ever heard the name of Garrideb?'

I admitted that I had not.

'Well, if you can lay your hands upon a Garrideb, there's money in it.'

'Why?'

'Ah, that's a long story—rather a whimsical one, too. I don't think in all our explorations of human complexities we have ever come upon anything more singular. The fellow will be here presently for cross-examination, so I won't open the matter up till he comes. But meanwhile, that's the name we want.'

The telephone directory* lay on the table beside me, and I turned over the pages in a rather hopeless quest. But to my amazement there was this strange name in its due place. I gave a cry of triumph.

'Here you are, Holmes! Here it is!'

Holmes took the book from my hand.

' "Garrideb, N.,' " he read, ' "136 Little Ryder Street, W." Sorry to disappoint you, my dear Watson, but this is the man himself. That is the address upon his letter. We want another to match him.'

Mrs Hudson had come in with a card upon a tray. I took it up and glanced at it.

'Why, here it is!' I cried in amazement. 'This is a different initial. John Garrideb, Counsellor at Law,* Moorville,* Kansas, USA.'

Holmes smiled as he looked at the card. 'I am afraid you must make yet another effort, Watson,' said he. 'This gentleman is also in the plot already, though I certainly did not expect to see him this morning. However, he is in a position to tell us a good deal which I want to know.'

A moment later he was in the room. Mr John Garrideb, Counsellor at Law, was a short, powerful man with the round, fresh, clean-shaven face characteristic of so many American men of affairs. The general effect was chubby and rather childlike, so that one received the impression of quite a young man with a broad set smile upon his face. His eyes, however, were arresting. Seldom in any human head have I seen a pair which bespoke a more intense inward life, so bright were they, so alert, so responsive to every change of thought. His accent was American, but was not accompanied by any eccentricity of speech.*

'Mr Holmes?' he asked, glancing from one to the other. 'Ah, yes! Your pictures are not unlike you, sir, if I may say so. I believe you have had a letter from my namesake, Mr Nathan Garrideb, have you not?'

'Pray sit down,' said Sherlock Holmes. 'We shall, I fancy, have a good deal to discuss.' He took up his sheets of foolscap. 'You are, of course, the Mr John Garrideb mentioned in this document. But surely you have been in England some time?'

'Why do you say that, Mr Holmes?' I seemed to read sudden suspicion in those expressive eyes.

'Your whole outfit is English.'

Mr Garrideb forced a laugh. 'I've read of your tricks, Mr Holmes, but I never thought I would be the subject of them. Where do you read that?'

'The shoulder cut of your coat, the toes of your boots—could anyone doubt it?'

'Well, well, I had no idea I was so obvious a Britisher. But business brought me over here some time ago, and so, as you say, my outfit is nearly all London. However, I guess your time is of value, and we did not meet to talk about the cut of my socks. What about getting down to that paper you hold in your hand?'

Holmes had in some way ruffled our visitor, whose chubby face had assumed a far less amiable expression.

'Patience! Patience, Mr Garrideb!' said my friend in a soothing voice. 'Dr Watson would tell you that these little digressions of mine sometimes prove in the end to have some bearing on the matter. But why did Mr Nathan Garrideb not come with you?'

'Why did he ever drag you into it at all?' asked our visitor, with a sudden outflame of anger. 'What in thunder had you to do with it? Here was a bit of professional business between two gentlemen, and one of them must needs call in a detective! I saw him this morning, and he told me this fool trick* he had played me, and that's why I am here. But I feel bad about it, all the same.'

'There was no reflection upon you, Mr Garrideb. It was simply zeal upon his part to gain your end—an end which is, I understand, equally vital for both of you. He knew that I had means of getting information, and, therefore, it was very natural that he should apply to me.'

Our visitor's angry face gradually cleared.

'Well, that puts it different,'* said he. 'When I went to see him this morning and he told me he had sent to a detective, I just asked for your address and came right away. I don't want police butting into a private matter. But if you are content just to help us find the man, there can be no harm in that.'

'Well, that is just how it stands,' said Holmes. 'And now, sir, since you are here, we had best have a clear account from your own lips. My friend here knows nothing of the details.'

Mr Garrideb surveyed me with not too friendly a gaze.

'Need he know?' he asked.

'We usually work together.'

'Well, there's no reason it should be kept a secret. I'll give you the facts as short as I can make them. If you came from Kansas I would not need to explain to you who Alexander Hamilton Garrideb* was. He made his money in real estate, and afterwards in the wheat pit* at Chicago, but he spent it in buying up as much land as would make one of your counties, lying along the Arkansas River,* west of Fort Dodge.* It's grazing-land and lumber-land* and arable land and mineralized land, and just every sort of land that brings dollars to the man that owns it.

'He had no kith nor kin—or, if he had, I never heard of it. But he took a kind of pride in the queerness of his name. That was what brought us together. I was in the law at Topeka,* and one day I had a visit from the old man, and he was tickled to death to meet another man with his own name. It was his pet fad, and he was dead set to find out if there were any more Garridebs in the world. "Find me another!" said he. I told him I was a busy man and could not spend my life hiking round the world in search of Garridebs. "None the less," said he, "that is just what you will do if things pan out* as I planned them." I thought he was joking, but there was a powerful lot of meaning in the words, as I was soon to discover.

'For he died within a year of saying them, and he left a will behind him. It was the queerest will that has ever been filed in the State of Kansas. His property was divided into three parts, and I was to have one in condition that I found two Garridebs who would share the remainder. It's five million dollars for each if it is a cent, but we can't lay a finger on it until we all three stand in a row.

'It was so big a chance that I just let my legal practice slide and I set forth looking for Garridebs. There is not one in the United States. I went through it, sir, with a fine-toothed comb and never a Garrideb could I catch. Then I tried the old country. Sure enough there was the name in the London Telephone Directory. I went after him two days ago and explained the whole matter to him. But he is a lone man, like myself, with some women relations, but no men. It says three adult men in the will. So you see we still have a vacancy,* and if you can help to fill it we will be very ready to pay your charges.'

'Well, Watson,' said Holmes, with a smile, 'I said it was rather whimsical, did I not? I should have thought, sir, that your obvious way was to advertise in the agony columns* of the papers.'

'I have done that, Mr Holmes. No replies.'

'Dear me! Well, it is certainly a most curious little problem. I may take a glance at it in my leisure. By the way, it is curious that you should have come from Topeka. I used to have a correspondent—he is dead now—old Dr Lysander Starr,* who was Mayor in 1890.'

'Good old Dr Starr!' said our visitor. 'His name is still honoured. Well, Mr Holmes, I suppose all we can do is to report to you and let you know how we progress. I reckon you will hear within a day or two.' With this assurance our American bowed and departed.

Holmes had lit his pipe, and he sat for some time with a curious smile upon his face.

'Well?' I asked at last.

'I am wondering, Watson—just wondering!'

'At what?'

Holmes took his pipe from his lips.

'I am wondering, Watson, what on earth could be the object of this man in telling us such a rigmarole of lies. I nearly asked him so—for there are times when a brutal frontal attack is the best policy—but I judged it better to let him think he had fooled us. Here is a man with an English coat frayed at the elbow and trousers bagged at the knee with a year's wear, and yet by this document and by his own account he is a provincial American lately landed in

London. There have been no advertisements in the agony columns. You know that I miss nothing there. They are my favourite covert for putting up a bird,* and I would never have overlooked such a cock pheasant as that. I never knew a Dr Lysander Starr of Topeka. Touch him where you would he was false. I think the fellow is really an American, but he has worn his accent smooth with years of London. What is his game, then, and what motive lies behind this preposterous search for Garridebs? It's worth our attention, for, granting that the man is a rascal, he is certainly a complex and ingenious one. We must now find out if our other correspondent is a fraud also. Just ring him up, Watson.'

I did so, and heard a thin, quavering voice at the other end of the line.

'Yes, yes, I am Mr Nathan Garrideb. Is Mr Holmes there? I should very much like to have a word with Mr Holmes.'

My friend took the instrument and I heard the usual syncopated* dialogue.

'Yes, he has been here. I understand that you don't know him . . . How long? . . . Only two days! . . . Yes, yes, of course, it is a most captivating prospect. Will you be at home this evening? I suppose your namesake will not be there? . . . Very good, we will come then, for I would rather have a chat without him . . . Dr Watson will come with me . . . I understood from your note that you did not go out often . . . Well, we shall be round about six. You need not mention it to the American lawyer . . . Very good. Good-bye!'

It was twilight of a lovely spring evening,* and even Little Ryder Street, one of the smaller offshoots from the Edgware Road, within a stone-cast of old Tyburn Tree* of evil memory, looked golden and wonderful in the slanting rays of the setting sun. The particular house to which we were directed was a large, old-fashioned, Early Georgian* edifice with a flat brick face broken only by two deep bay windows on the ground floor. It was on this ground floor that our client lived, and, indeed, the low windows proved to be the front of the huge room in which he spent his waking hours. Holmes pointed as we passed to the small brass plate which bore the curious name.

'Up some years, Watson,' he remarked, indicating its discoloured surface. 'It's *his* real name, anyhow, and that is something to note.'

The house had a common stair, and there were a number of names painted in the hall, some indicating offices and some private chambers. It was not a collection of residential flats, but rather the abode of Bohemian bachelors. Our client opened the door for us himself and apologized by saying that the woman in charge left at four o'clock. Mr Nathan Garrideb proved to be a very tall loose-jointed, round-backed person, gaunt and bald, some sixty-odd years of age. He had a cadaverous* face, with the dull dead skin of a man to whom exercise was unknown. Large round spectacles and a small projecting goat's beard combined with his stooping attitude to give him an expression of peering curiosity. The general effect, however, was amiable, though eccentric.

The room was as curious as its occupant. It looked like a small museum. It was both broad and deep, with cupboards and cabinets all round, crowded with specimens, geological and anatomical. Cases of butterflies and moths flanked each side of the entrance. A large table in the centre was littered with all sorts of debris, while the tall brass tube of a powerful microscope bristled up amongst them. As I glanced round I was surprised at the universality of the man's interests. Here was a case of ancient coins. There was a cabinet of flint instruments.* Behind his central table was a large cupboard of fossil bones. Above was a line of plaster skulls with such names as 'Neanderthal', 'Heidelberg', 'Cromagnon'* printed beneath them. It was clear that he was a student of many subjects. As he stood in front of us now, he held a piece of chamois leather in his right hand with which he was polishing a coin.

'Syracusan*—of the best period,' he explained, holding it up. 'They degenerated greatly towards the end. At their best

I hold them supreme, though some prefer the Alexandrian school.* You will find a chair here, Mr Holmes. Pray allow me to clear these bones. And you, sir—ah, yes, Dr Watson—if you would have the goodness to put the Japanese vase to one side. You see round me my little interests in life. My doctor lectures me about never going out, but why should I go out when I have so much to hold me here? I can assure you that the adequate cataloguing of one of those cabinets would take me three good months.'

Holmes looked round him with curiosity.

'But do you tell me that you *never* go out?' he said.

'Now and again I drive down to Sotheby's or Christie's.* Otherwise I very seldom leave my room. I am not too strong, and my researches are very absorbing. But you can imagine, Mr Holmes, what a terrific shock—pleasant but terrific—it was for me when I heard of this unparalleled good fortune. It only needs one more Garrideb to complete the matter, and surely we can find one. I had a brother, but he is dead, and female relatives are disqualified. But there must surely be others in the world. I had heard that you handled strange cases, and that was why I sent to you. Of course, this American gentleman is quite right, and I should have taken his advice first, but I acted for the best.'

'I think you acted very wisely indeed,' said Holmes. 'But are you really anxious to acquire an estate in America?'

'Certainly not, sir. Nothing would induce me to leave my collection. But this gentleman has assured me that he will buy me out as soon as we have established our claim. Five million dollars was the sum named. There are a dozen specimens in the market at the present moment which fill gaps in my collection, and which I am unable to purchase for want of a few hundred pounds. Just think what I could do with five million dollars. Why, I have the nucleus of a national collection. I shall be the Hans Sloane* of my age.'

His eyes gleamed behind his spectacles. It was very clear that no pains would be spared by Mr Nathan Garrideb in finding a namesake.

'I merely called to make your acquaintance, and there is no reason why I should interrupt your studies,' said Holmes. 'I prefer to establish personal touch with those with whom I do business. There are few questions I need ask, for I have your very clear narrative in my pocket, and I filled up the blanks when this American gentleman called. I understand that up to this week you were unaware of his existence.'

'That is so. He called last Tuesday.'

'Did he tell you of our interview to-day?'

'Yes, he came straight back to me. He had been very angry.'

'Why should he be angry?'

'He seemed to think it was some reflection on his honour. But he was quite cheerful again when he returned.'

'Did he suggest any course of action?'

'No, sir, he did not.'

'Has he had, or asked for, any money from you?'

'No, sir, never!'

'You see no possible object he has in view?'

'None, except what he states.'

'Did you tell him of our telephone appointment?'*

'Yes, sir, I did.'

Holmes was lost in thought. I could see that he was puzzled.

'Have you any articles of great value in your collection?'

'No, sir. I am not a rich man. It is a good collection, but not a very valuable one.'

'You have no fear of burglars?'

'Not the least.'

'How long have you been in these rooms?'

'Nearly five years.'

Holmes's cross-examination was interrupted by an imperative knocking at the door. No sooner had our client unlatched it than the American lawyer burst excitedly into the room.

'Here you are!' he cried, waving a paper over his head. 'I thought I should be in time to get you. Mr Nathan Garrideb, my congratulations! You are a rich man, sir. Our business is happily finished and all is well. As to you, Mr Holmes, we can only say we are sorry if we have given you any useless trouble.'

He handed over the paper to our client, who stood staring at a marked advertisement. Holmes and I leaned forward and read it over his shoulder. This is how it ran:

HOWARD GARRIDEB

Constructor of Agricultural Machinery

Binders, reapers' steam and hand plows, drills,* harrows,* farmers' carts, buckboards,* and all other appliances

Estimates for Artesian Wells*

Apply Grosvenor Buildings, Aston*

'Glorious!' gasped our host. 'That makes our third man.'

'I had opened up inquiries in Birmingham,'* said the American, 'and my agent there has sent me this advertisement from a local paper. We must hustle and put the thing through. I have written to this man and told him that you will see him in his office to-morrow afternoon at four o'clock.'

'You want *me* to see him?'

'What do you say, Mr Holmes? Don't you think it would be wiser? Here am I, a wandering American with a wonderful tale. Why should he believe what I tell him? But you are a Britisher with solid references, and he is bound to take notice of what you say. I would go with you if you wished, but I have a very busy day to-morrow, and I could always follow you if you are in any trouble.'

'Well, I have not made such a journey for years.'

'It is nothing, Mr Garrideb. I have figured out* your connections.* You leave at twelve and should be there soon after two. Then you can be back the same night. All you have to do is to see this man, explain the matter, and get an affidavit of his existence. By the Lord!' he added hotly, 'considering I've come all the way from the centre of

America, it is surely little enough if you go a hundred miles in order to put this matter through.'

'Quite so,' said Holmes. 'I think what this gentleman says is very true.'

Mr Nathan Garrideb shrugged his shoulders with a disconsolate air. 'Well, if you insist I shall go,' said he. 'It is certainly hard for me to refuse you anything, considering the glory of hope that you have brought into my life.'

'Then that is agreed,' said Holmes, 'and no doubt you will let me have a report as soon as you can.'

'I'll see to that,' said the American. 'Well,' he added, looking at his watch, 'I'll have to get on. I'll call to-morrow, Mr Nathan, and see you off to Birmingham. Coming my way, Mr Holmes? Well, then, good-bye, and we may have good news for you to-morrow night.'

I noticed that my friend's face cleared when the American left the room, and the look of thoughtful perplexity had vanished.

'I wish I could look over your collection, Mr Garrideb,' said he. 'In my profession all sorts of odd knowledge comes useful, and this room of yours is a storehouse of it.'

Our client shone with pleasure and his eyes gleamed from behind his big glasses.

'I had always heard, sir, that you were a very intelligent man,' said he. 'I could take you round now, if you have the time.'

'Unfortunately, I have not. But these specimens are so well labelled and classified that they hardly need your personal explanation. If I should be able to look in to-morrow, I presume that there would be no objection to my glancing over them?'

'None at all. You are most welcome. The place will, of course, be shut up, but Mrs Saunders is in the basement up to four o'clock and would let you in with her key.'

'Well, I happen to be clear to-morrow afternoon. If you would say a word to Mrs Saunders it would be quite in order. By the way, who is your house-agent?'

Our client was amazed at the sudden question.

'Holloway and Steele, in the Edgware Road. But why?'

'I am a bit of an archaeologist myself when it comes to houses,' said Holmes, laughing. 'I was wondering if this was Queen Anne* or Georgian.'

'Georgian, beyond doubt.'

'Really. I should have thought a little earlier. However, it is easily ascertained. Well, good-bye, Mr Garrideb, and may you have every success in your Birmingham journey.'

The house-agent's was close by, but we found that it was closed for the day, so we made our way back to Baker Street. It was not till after dinner that Holmes reverted to the subject.

'Our little problem draws to a close,' said he. 'No doubt you have outlined the solution in your own mind.'

'I can make neither head nor tail of it.'

'The head is surely clear enough and the tail we should see to-morrow. Did you notice nothing curious about that advertisement?'

'I saw that the word "plough" was mis-spelt.'

'Oh, you did notice that, did you? Come, Watson, you improve all the time. Yes, it was bad English but good American. The printer had set it up as received. Then the buckboards. That is American also. And artesian wells are commoner with them than with us. It was a typical American advertisement, but purporting to be from an English firm. What do you make of that?'

'I can only suppose that this American lawyer put it in himself. What his object was I fail to understand.'

'Well, there are alternative explanations. Anyhow, he wanted to get this good old fossil up to Birmingham. That is very clear. I might have told him that he was clearly going on a wild-goose chase, but, on second thoughts, it seemed better to clear the stage by letting him go. To-morrow, Watson—well, to-morrow will speak for itself.'

Holmes was up and out early. When he returned at lunch-time I noticed that his face was very grave.

'This is a more serious matter than I had expected, Watson,' said he. 'It is fair to tell you so, though I know it will be only an additional reason to you for running your head into danger. I should know my Watson by now. But there *is* danger, and you should know it.'

'Well, it is not the first we have shared, Holmes. I hope it may not be the last. What is the particular danger this time?'

'We are up against a very hard case. I have identified Mr John Garrideb, Counsellor at Law. He is none other than "Killer" Evans, of sinister and murderous reputation.'

'I fear I am none the wiser.'

'Ah, it is not part of your profession to carry about a portable Newgate Calendar* in your memory. I have been down to see friend Lestrade* at the Yard.* There may be an occasional want of imaginative intuition down there, but they lead the world for thoroughness and method. I had an idea that we might get on the track of our American friend in their records. Sure enough, I found his chubby face smiling up at me from the Rogues' Portrait Gallery.* James Winter, *alias* Morecroft, *alias* Killer Evans, was the inscription below.' Holmes drew an envelope from his pocket. 'I scribbled down a few points from his dossier. Aged forty-six. Native of Chicago. Known to have shot three men in the States. Escaped from penitentiary through political influence. Came to London in 1893. Shot a man over cards in a night-club in the Waterloo Road in January, 1895. Man died, but he was shown to have been the aggressor in the row. Dead man was identified as Rodger Prescott,* famous as forger and coiner in Chicago. Killer Evans released in 1901. Has been under police supervision since, but so far as known has led an honest life. Very dangerous man, usually carries arms and is prepared to use them. That is our bird, Watson—a sporting bird, as you must admit.'

'But what is his game?'

'Well, it begins to define itself. I have been to the house-agents. Our client, as he told us, has been there five years. It was unlet for a year before then. The previous tenant was a gentleman at large named Waldron. Waldron's appearance was well remembered at the office. He had suddenly vanished and nothing more been heard of him. He was a tall, bearded man with very dark features. Now, Prescott, the man whom Killer Evans had shot, was, according to Scotland Yard, a tall, dark man with a beard. As a working hypothesis, I think we may take it that Prescott, the American criminal, used to live in the very room which our innocent friend now devotes to his museum. So at last we get a link, you see.'

'And the next link?'

'Well, we must go now and look for that.'

He took a revolver from the drawer and handed it to me. 'I have my old favourite with me. If our Wild West friend tries to live up to his nickname, we must be ready for him. I'll give you an hour for a siesta, Watson, and then I think it will be time for our Ryder Street adventure.'

It was just four o'clock when we reached the curious apartment of Nathan Garrideb. Mrs Saunders, the caretaker, was about to leave, but she had no hesitation in admitting us, for the door shut with a spring lock and Holmes promised to see that all was safe before we left. Shortly afterwards the outer door closed, her bonnet passed the bow window, and we knew that we were alone in the lower floor of the house. Holmes made a rapid examination of the premises. There was one cupboard in a dark corner which stood out a little from the wall. It was behind this that we eventually crouched, while Holmes in a whisper outlined his intentions.

'He wanted to get our amiable friend out of his room—that is very clear, and, as the collector never went out, it took some planning to do it. The whole of this Garrideb invention was apparently for no other end. I must say, Watson, that there is a certain devilish ingenuity* about it, even if the queer name of the tenant did give him an opening which he could hardly have expected. He wove his plot with remarkable cunning.'

'But what did he want?'

'Well, that is what we are here to find out. It has nothing whatever to do with our client, so far as I can read the situation. It is something connected with the man he murdered—the man who may have been his confederate in crime. There is some guilty secret in the room. That is how I read it. At first I thought our friend might have something in his collection more valuable than he knew—something worth the attention of a big criminal. But the fact that Rodger Prescott of evil memory inhabited these rooms points to some deeper reason. Well, Watson, we can but possess our souls in patience and see what the hour may bring.'

That hour was not long in striking. We crouched closer in the shadow as we heard the outer door open and shut. Then came the sharp, metallic snap of a key, and the American was in the room. He closed the door softly behind him, took a sharp glance around him to see that all was safe, threw off his overcoat, and walked up to the central table with the brisk manner of one who knows exactly what he has to do and how to do it. He pushed the table to one side, tore up the square of carpet on which it rested, rolled it completely back, and then, drawing a jemmy* from his inside pocket, he knelt down and worked vigorously upon the floor. Presently we heard the sound of sliding boards, and an instant later a square had opened in the planks. Killer Evans struck a match, lit a stump of candle, and vanished from our view.

Clearly our moment had come. Holmes touched my wrist as a signal, and together we stole across to the open trap-door. Gently as we moved, however, the old floor must have creaked under our feet, for the head of our American, peering anxiously round, emerged suddenly from the open space. His face turned upon us with a glare of baffled rage, which gradually softened into a rather shamefaced grin as he realized that two pistols were pointed at his head.

'Well, well!' said he, coolly, as he scrambled to the surface. 'I guess you have been one too many for me, Mr Holmes. Saw through my game, I suppose, and played me for a sucker* from the first. Well, sir, I hand it to you; you have me beat and—'

In an instant he had whisked out a revolver from his breast and had fired two shots. I felt a sudden hot sear as if a red-hot iron had been pressed to my thigh. There was a crash as Holmes's pistol came down on the man's head. I had a vision of him sprawling upon the floor with blood running down his face while Holmes rummaged him for weapons. Then my friend's wiry arms were round me and he was leading me to a chair.

'You're not hurt, Watson? For God's sake, say that you are not hurt!'

It was worth a wound—it was worth many wounds—to know the depth of loyalty and love which lay behind that cold mask. The clear, hard eyes were dimmed for a moment, and the firm lips were shaking. For the one and only time I caught a glimpse of a great heart as well as of a great brain. All my years of humble but single-minded service culminated in that moment of revelation.

'It's nothing, Holmes. It's a mere scratch.'

He had ripped up my trousers with his pocket-knife.

'You are right,' he cried, with an immense sigh of relief. 'It is quite superficial.' His face set like flint as he glared at our prisoner, who was sitting up with a dazed face. 'By the Lord, it is as well for you. If you had killed Watson, you would not have got out of this room alive. Now, sir, what have you to say for yourself?'

He had nothing to say for himself. He only lay and scowled. I leaned on Holmes's arm, and together we looked down into the small cellar which had been disclosed by the secret flap. It was still illuminated by the candle which Evans had taken down with him. Our eyes fell upon a mass of rusted machinery, great rolls of paper, a litter of bottles, and, neatly arranged upon a small table, a number of neat little bundles.

'A printing press—a counterfeiter's outfit,' said Holmes.

'Yes, sir,' said our prisoner, staggering slowly to his feet and then sinking into the chair. 'The greatest counterfeiter London ever saw. That's Prescott's machine, and those bundles on the table are two thousand of Prescott's notes worth a hundred each and fit to pass anywhere. Help yourselves, gentlemen. Call it a deal and let me beat it.'

Holmes laughed.

'We don't do things like that, Mr Evans. There is no bolt-hole for you in this country. You shot this man Prescott, did you not?'

'Yes, sir, and got five years for it, though it was he who pulled on me. Five years—when I should have had a medal the size of a soup plate. No living man could tell a Prescott from a Bank of England, and if I hadn't put him out he would have flooded London with them. I was the only one in the world who knew where he made them. Can you wonder that I wanted to get to the place? And can you wonder that when I found this crazy boob* of a bug-hunter with the queer name squatting right on the top of it, and never quitting his room, I had to do the best I could to shift him? Maybe I would have been wiser if I had put him away. It would have been easy enough, but I'm a soft-hearted guy that can't begin shooting unless the other man has a gun also. But say, Mr Holmes, what have I done wrong, anyhow? I've not used this plant. I've not hurt this old stiff. Where do you get me?'*

'Only attempted murder, so far as I can see,' said Holmes. 'But that's not our job. They take that at the next stage. What we wanted at present was just your sweet self. Please give the Yard a call, Watson. It won't be entirely unexpected.'

So those were the facts about Killer Evans and his remarkable invention of the three Garridebs. We heard later that our poor old friend never got over the shock of his dissipated dreams. When his castle in the air fell down, it buried him beneath the ruins. He was last heard of at a nursing-home in Brixton.* It was a glad day at the Yard when the Prescott outfit was discovered, for, though they knew that it existed, they had never been able, after the death of the man, to find out where it was. Evans had indeed done great service, and caused several worthy CID men to sleep the sounder, for the counterfeiter stands in a class by himself as a public danger. They would willingly have subscribed to that soup-plate medal of which the criminal had spoken, but an unappreciative Bench* took a less favourable view, and the Killer returned to those shades from which he had just emerged.*

The Illustrious Client

'IT can't hurt now,' was Mr Sherlock Holmes's comment when, for the tenth time in as many years, I asked his leave to reveal the following narrative. So it was that at last I obtained permission to put on record what was, in some ways, the supreme moment of my friend's career.*

Both Holmes and I had a weakness for the Turkish Bath.* It was over a smoke in the pleasant lassitude of the drying-room that I found him less reticent and more human than anywhere else. On the upper floor of the Northumberland Avenue* establishment there is an isolated corner where two couches lie side by side, and it was on these that we lay upon September 3,* 1902, the day when my narrative begins. I had asked him whether anything was stirring, and for answer he had shot his long, thin, nervous arm out of the sheets which enveloped him and had drawn an envelope from the inside pocket of the coat which hung beside him.

'It may be some fussy, self-important fool, it may be a matter of life or death,' said he, as he handed me the note. 'I know no more than this message tells me.'

It was from the Carlton Club,* and dated the evening before. This is what I read:

Sir James Damery presents his compliments to Mr Sherlock Holmes, and will call upon him at 4.30 to-morrow. Sir James begs to say that the matter upon which he desires to consult Mr Holmes is very delicate, and also very important. He trusts, therefore, that Mr Holmes will make every effort to grant this interview, and that he will confirm it over the telephone to the Carlton Club.

'I need not say that I have confirmed it, Watson,' said Holmes, as I returned the paper. 'Do you know anything of this man Damery?'

'Only that his name is a household word in Society.'

'Well, I can tell you a little more than that. He has rather a reputation for arranging delicate matters which are to be kept out of the papers. You may remember his negotiations with Sir George Lewis* over the Hammerford Will case. He is a man of the world with a natural turn for diplomacy. I am bound, therefore, to hope that it is not a false scent and that he has some real need for our assistance.'

'Our?'

'Well, if you will be so good, Watson.'

'I shall be honoured.'

'Then you have the hour—four-thirty. Until then we can put the matter out of our heads.'

I was living in my own rooms in Queen Anne Street at the time, but I was round at Baker Street* before the time named. Sharp to the half-hour, Colonel Sir James Damery* was announced. It is hardly necessary to describe him, for many will remember that large, bluff, honest personality, that broad, clean-shaven face, above all, that pleasant, mellow voice. Frankness shone from his grey Irish eyes, and good humour played round his mobile, smiling lips. His lucent top-hat, his dark frock-coat, indeed, every detail, from the pearl pin in the black satin cravat* to the lavender spats* over the varnished shoes, spoke of the meticulous care in dress for which he was famous. The big, masterful aristocrat dominated the little room.

'Of course, I was prepared to find Dr Watson,' he remarked, with a courteous bow. 'His collaboration may be very necessary, for we are dealing on this occasion, Mr Holmes, with a man to whom violence is familiar and who will, literally, stick at nothing. I should say that there is no more dangerous man in Europe.'*

'I have had several opponents to whom that flattering term has been applied,' said Holmes, with a smile. 'Don't you smoke? Then you will excuse me if I light my pipe. If your man is more dangerous than the late Professor Moriarty,* or than the living Colonel Sebastian Moran,* then he is indeed worth meeting. May I ask his name?'

'Have you ever heard of Baron Gruner?'

'You mean the Austrian murderer?'

Colonel Damery threw up his kid-gloved* hands with a laugh. 'There is no getting past you, Mr Holmes! Wonderful! So you have already sized him up as a murderer?'

'It is my business to follow the details of Continental crime. Who could possibly have read what happened at Prague* and have any doubts as to the man's guilt! It was a purely technical legal point and the suspicious death of a witness that saved him! I am as sure that he killed his wife when the so-called "accident" happened in the Splügen Pass* as if I had seen him do it. I knew, also, that he had come to England, and had a presentiment that sooner or later he would find me some work to do. Well, what has Baron Gruner been up to? I presume it is not this old tragedy which has come up again?'

'No, it is more serious than that. To revenge crime is important, but to prevent it is more so. It is a terrible thing, Mr Holmes, to see a dreadful event, an atrocious situation, preparing itself before your eyes, to clearly understand whither it will lead, and yet to be utterly unable to avert it. Can a human being be placed in a more trying position?'

'Perhaps not.'

'Then you will sympathize with the client in whose interests I am acting.'

'I did not understand that you were merely an intermediary. Who is the principal?'

'Mr Holmes, I must beg you not to press that question. It is important that I should be able to assure him that his honoured name has been in no way dragged into the matter. His motives are, to the last degree, honourable and chivalrous, but he prefers to remain unknown. I need not say that your fees will be assured and that you will be given a perfectly free hand. Surely the actual name of your client is immaterial?'

'I am sorry,' said Holmes. 'I am accustomed to have mystery at one end of my cases, but to have it at both ends is too confusing. I fear, Sir James, that I must decline to act.'

Our visitor was greatly disturbed. His large, sensitive face was darkened with emotion and disappointment.

'You hardly realize the effect of your own action, Mr Holmes,' said he. 'You place me in a most serious dilemma, for I am perfectly certain that you would be proud to take over the case if I could give you the facts, and yet a promise forbids me from revealing them all. May I, at least, lay all that I can before you?'

'By all means, so long as it is understood that I commit myself to nothing.'

'That is understood. In the first place, you have no doubt heard of General de Merville?'

'De Merville of Khyber* fame? Yes, I have heard of him.'

'He has a daughter, Violet de Merville, young, rich, beautiful, accomplished, a wonder-woman in every way. It is this daughter, this lovely, innocent girl, whom we are endeavouring to save from the clutches of a fiend.'

'Baron Gruner has some hold over her, then?'

'The strongest of all holds where a woman is concerned—the hold of love. The fellow is, as you may have heard, extraordinarily handsome, with a most fascinating manner, a gentle voice, and that air of romance and mystery which means so much to a woman. He is said to have the whole sex at his mercy and to have made ample use of the fact.'

'But how came such a man to meet a lady of the standing of Miss Violet de Merville?'

'It was on a Mediterranean yachting voyage. The company, though select, paid their own passages. No doubt the promoters hardly realized the Baron's true character until it was too late. The villain attached himself to the lady, and with such effect that he has completely and absolutely won her heart. To say that she loves him hardly expresses it. She dotes upon him, she is obsessed by him. Outside of him there is nothing on earth. She will not hear one word against him. Everything has been done to cure her of her madness, but in vain. To sum up, she proposes to marry him next month. As she is of age and has a will of iron, it is hard to know how to prevent her.'

'Does she know about the Austrian episode?'

'The cunning devil has told her every unsavoury public scandal of his past life, but always in such a way as to make himself out to be an innocent martyr. She absolutely accepts his version and will listen to no other.'

'Dear me! But surely you have inadvertently let out the name of your client? It is no doubt General de Merville.'

Our visitor fidgeted in his chair.

'I could deceive you by saying so, Mr Holmes, but it would not be true. De Merville is a broken man. The strong soldier has been utterly demoralized by this incident. He has lost the nerve which never failed him on the battlefield and has become a weak, doddering old man, utterly incapable of contending with a brilliant, forceful rascal like this Austrian. My client, however, is an old friend, one who has known the General intimately for many years and taken a paternal interest in this young girl since she wore short frocks. He cannot see this tragedy consummated without some attempt to stop it. There is nothing in which Scotland Yard can act. It was his own suggestion that you should be called in, but it was, as I have said, on the express stipulation that he should not be personally involved in the matter. I have no doubt, Mr Holmes, with your great powers you could easily trace my client back through me, but I must ask you, as a point of honour, to refrain from doing so, and not to break in upon his incognito.'

Holmes gave a whimsical smile.

'I think I may safely promise that,' said he. 'I may add that your problem interests me, and that I shall be prepared to look into it. How shall I keep in touch with you?'

'The Carlton Club will find me. But, in case of emergency, there is a private telephone call, "XX.31." '

Holmes noted it down and sat, still smiling, with the open memorandum-book upon his knee.

'The Baron's present address, please?'

'Vernon Lodge, near Kingston.* It is a large house. He has been fortunate in some rather shady speculations and is a rich man, which, naturally, makes him a more dangerous antagonist.'

'Is he at home at present?'

'Yes.'

'Apart from what you have told me, can you give me any further information about the man?'

'He has expensive tastes. He is a horse fancier. For a short time he played polo at Hurlingham,* but then this Prague affair got noised about and he had to leave. He collects books and pictures. He is a man with a considerable artistic side to his nature. He is, I believe, a recognized authority upon Chinese pottery, and has written a book upon the subject.'

'A complex mind,' said Holmes. 'All great criminals have that. My old friend Charlie Peace* was a violin virtuoso. Wainwright* was no mean artist. I could quote many more. Well, Sir James, you will inform your client that I am turning my mind upon Baron Gruner. I can say no more. I have some sources of information of my own, and I dare say we may find some means of opening the matter up.'

When our visitor had left us, Holmes sat so long in deep thought that it seemed to me that he had forgotten my presence. At last, however, he came briskly back to earth.

'Well, Watson, any views?' he asked.

'I should think you had better see the young lady herself.'

'My dear Watson, if her poor old broken father cannot move her, how shall I, a stranger, prevail? And yet there is something in the suggestion if all else fails. But I think we must begin from a different angle. I rather fancy that Shinwell Johnson* might be a help.'

I have not had occasion to mention Shinwell Johnson in these memoirs because I have seldom drawn my cases from the latter phases of my friend's career. During the first years of the century he became a valuable assistant. Johnson, I grieve to say, made his name first as a very dangerous villain and served two terms at Parkhurst.* Finally, he repented and allied himself to Holmes, acting as his agent in the huge criminal underworld of London, and obtaining information which often proved to be of vital importance. Had Johnson been a 'nark'* of the police he would soon have been exposed, but as he dealt with cases which never came directly into the courts, his activities were never realized by his companions. With the glamour of his two convictions upon him, he had the *entrée* of every night-club, doss-house, and gambling-den in the town, and his quick observation and active brain made him an ideal agent for gaining information. It was to him that Sherlock Holmes now proposed to turn.

It was not possible for me to follow the immediate steps taken by my friend, for I had some pressing professional business of my own, but I met him by appointment that evening at Simpson's,* where, sitting at a small table in the front window, and looking down at the rushing stream of life in the Strand,* he told me something of what had passed.

'Johnson is on the prowl,' said he. 'He may pick up some garbage in the darker recesses of the underworld, for it is down there, amid the black roots of crime, that we must hunt for this man's secrets.'

'But, if the lady will not accept what is already known, why should any fresh discovery of yours turn her from her purpose?'

'Who knows, Watson? Woman's heart and mind are insoluble puzzles to the male. Murder might be condoned or explained, and yet some smaller offence might rankle. Baron Gruner remarked to me—'

'He remarked to you!'

'Oh, to be sure, I had not told you of my plans! Well, Watson, I love to come to close grips with my man. I like to meet him eye to eye and read for myself the stuff that he is made of. When I had given Johnson his instructions, I took a cab out to Kingston and found the Baron in a most affable mood.'

'Did he recognize you?'

'There was no difficulty about that, for I simply sent in my card. He is an excellent antagonist, cool as ice, silky voiced and soothing as one of your fashionable consultants, and poisonous as a cobra. He has breed in him, a real aristocrat of crime, with a superficial suggestion of afternoon-tea and all the cruelty of the grave behind it. Yes, I am glad to have had my attention called to Baron Adelbert Gruner.'

'You say he was affable?'

'A purring cat who thinks he sees prospective mice. Some people's affability is more deadly than the violence of coarser souls. His greeting was characteristic. "I rather thought I should see you sooner or later, Mr Holmes," said he. "You have been engaged, no doubt, by General de Merville to endeavour to stop my marriage with his daughter, Violet. That is so, is it not?"

'I acquiesced.

' "My dear man," said he, "you will only ruin your own well-deserved reputation. It is not a case in which you can possibly succeed. You will have barren work, to say nothing of incurring some danger. Let me very strongly advise you to draw off at once."

' "It is curious," I answered, "but that was the very advice which I had intended to give you. I have a respect for your brains, Baron, and the little which I have seen of your personality has not lessened it. Let me put it to you as man to man. No one wants to rake up your past and make you unduly uncomfortable. It is over, and you are now in smooth waters, but if you persist in this marriage you will raise up a swarm of powerful enemies who will never leave you alone until they have made England too hot to hold you. Is the game worth it? Surely you would be wiser if you left the lady alone. It would not be pleasant for you if these facts of your past were brought to her notice."

'The Baron has little waxed tips of hair under his nose, like the short antennae of an insect. These quivered with amusement as he listened, and he finally broke into a gentle chuckle.

' "Excuse my amusement, Mr Holmes," said he, "but it is really funny to see you trying to play a hand with no cards in it. I don't think anyone could do it better, but it is rather pathetic, all the same. Not a colour card there, Mr Holmes, nothing but the smallest of the small."

' "So you think."

' "So I know. Let me make the thing clear to you, for my own hand is so strong that I can afford to show it. I have been fortunate enough to win the entire affection of this lady. This was given to me in spite of the fact that I told her very clearly of all the unhappy incidents in my past life. I also told her that certain wicked and designing persons—I hope you recognize yourself—would come to her and tell her these things, and I warned her how to treat them. You have heard of post-hypnotic suggestion, Mr Holmes? Well, you will see how it works, for a man of personality can use hypnotism without any vulgar passes or tomfoolery. So she is ready for you and, I have no doubt, would give you an appointment, for she is quite amenable to her father's will—save only in the one little matter."

'Well, Watson, there seemed to be no more to say, so I took my leave with as much cold dignity as I could summon, but, as I had my hand on the door-handle, he stopped me.

' "By the way, Mr Holmes," said he, "did you know Le Brun, the French agent?"

' "Yes," said I.

' "Do you know what befell him?"

' "I heard that he was beaten by some Apaches* in the Montmartre* district and crippled for life."

' "Quite true, Mr Holmes. By a curious coincidence he had been inquiring into my affairs only a week before. Don't do it, Mr Holmes; it's not a lucky thing to do. Several have found that out. My last word to you is, go your own way and let me go mine. Good-bye!"

'So there you are, Watson. You are up to date now.'

'The fellow seems dangerous.'

'Mighty dangerous. I disregard the blusterer, but this is the sort of man who says rather less than he means.'

'Must you interfere? Does it really matter if he marries the girl?'

'Considering that he undoubtedly murdered his last wife, I should say it mattered very much. Besides, the client! Well, well, we need not discuss that. When you have finished your coffee you had best come home with me, for the blithe Shinwell will be there with his report.'

We found him sure enough, a huge, coarse, red-faced, scorbutic* man, with a pair of vivid black eyes which were the only external sign of the very cunning mind within. It seems that he had dived down into what was peculiarly his kingdom, and beside him on the settee was a brand* which he had brought up in the shape of a slim, flame-like young woman with a pale, intense face, youthful, and yet so worn with sin and sorrow that one read the terrible years which had left their leprous* mark upon her.

'This is Miss Kitty Winter,' said Shinwell Johnson, waving his fat hand as an introduction. 'What she don't know—well, there, she'll speak for herself. Put my hand right on her, Mr Holmes, within an hour of your message.'

'I'm easy to find,' said the young woman. 'Hell, London, gets me every time. Same address for Porky Shinwell. We're old mates, Porky, you and I. But, by Cripes! there is another who ought to be down in a lower hell than we if there was any justice in the world! That is the man you are after, Mr Holmes.'

Holmes smiled. 'I gather we have your good wishes, Miss Winter.'

'If I can help to put him where he belongs, I'm yours to the rattle,'* said our visitor, with fierce energy. There was an intensity of hatred in her white, set face and her blazing eyes such as woman seldom and man never can attain. 'You needn't go into my past, Mr Holmes. That's neither here nor there. But what I am Adelbert Gruner made me. If I could pull him down!' She clutched frantically with her hands into the air. 'Oh, if I could pull him into the pit where he had pushed so many!'

'You know how the matter stands?'

'Porky Shinwell has been telling me. He's after some other poor fool and wants to marry her this time. You want to stop it. Well, you surely know enough about this devil to prevent any decent girl in her senses wanting to be in the same parish with him.'

'She is not in her senses. She is madly in love. She has been told all about him. She cares nothing.'

'Told about the murder?'

'Yes.'

'My Lord, she must have a nerve!'

'She puts them all down as slanders.'

'Couldn't you lay proofs before her silly eyes?'

'Well, can you help us do so?'

'Ain't I a proof myself? If I stood before her and told her how he used me—'

'Would you do this?'

'Would I? Would I not!'

'Well, it might be worth trying. But he has told her most of his sins and had pardon from her, and I understand she will not reopen the question.'

'I'll lay he didn't tell her all,' said Miss Winter. 'I caught a glimpse of one or two murders besides the one that made such a fuss. He would speak of someone in his velvet way and then look at me with a steady eye and say: "He died within a month." It wasn't hot air, either. But I took little notice—you see, I loved him myself at that time. Whatever he did went with me, same as with this poor fool! There was just one thing that shook me. Yes, by Cripes! if it had not been for his poisonous, lying tongue that explains and soothes, I'd have left him that very night. It's a book he has*—a brown leather book with a lock, and his arms in gold on the outside. I think he was a bit drunk that night, or he would not have shown it to me.'

'What was it, then?'

'I tell you, Mr Holmes, this man collects women, and takes a pride in his collection, as some men collect moths or butterflies. He had it all in that book. Snapshot photographs, names, details, everything about them. It was a beastly book—a book no man, even if he had come from the gutter, could have put together. But it was Adelbert

Gruner's book all the same. "Souls I have ruined." He could have put that on the outside if he had been so minded. However, that's neither here nor there, for the book would not serve you, and, if it would, you can't get it.'

'Where is it?'

'How can I tell you where it is now? It's more than a year since I left him. I know where he kept it then. He's a precise, tidy cat of a man in many of his ways, so maybe it is still in the pigeon-hole of the old bureau in the inner study. Do you know his house?'

'I've been in the study,' said Holmes.

'Have you, though? You haven't been slow on the job if you only started this morning. Maybe dear Adelbert has met his match this time. The outer study is the one with the Chinese crockery in it—big glass cupboard between the windows. Then behind his desk is the door that leads to the inner study—a small room where he keeps papers and things.'

'Is he not afraid of burglars?'

'Adelbert is no coward. His worst enemy couldn't say that of him. He can look after himself. There's a burglar alarm at night. Besides, what is there for a burglar—unless they got away with all this fancy crockery?'

'No good,' said Shinwell Johnson, with the decided voice of the expert. 'No fence wants stuff of that sort that you can neither melt nor sell.'

'Quite so,' said Holmes. 'Well, now, Miss Winter, if you would call here to-morrow evening at five, I would consider in the meanwhile whether your suggestion of seeing this lady personally may not be arranged. I am exceedingly obliged to you for your co-operation. I need not say that my clients will consider liberally—'

'None of that, Mr Holmes,' cried the young woman. 'I am not out for money. Let me see this man in the mud, and I've got all I worked for—in the mud with my foot on his cursed face. That's my price. I'm with you tomorrow or any other day so long as you are on his track. Porky here can tell you always where to find me.'

I did not see Holmes again until the following evening, when we dined once more at our Strand restaurant. He shrugged his shoulders when I asked him what luck he had had in his interview. Then he told the story, which I would repeat in this way. His hard, dry statement needs some little editing to soften it into the terms of real life.

'There was no difficulty at all about the appointment,' said Holmes, 'for the girl glories in showing abject filial obedience in all secondary things in an attempt to atone for her flagrant breach of it in her engagement. The General 'phoned that all was ready, and the fiery Miss W. turned up according to schedule, so that at half-past five a cab deposited us outside 104 Berkeley Square, where the old soldier resides—one of those awful grey London castles which would make a church seem frivolous. A footman showed us into a great yellow-curtained drawing-room, and there was the lady awaiting us, demure, pale, self-contained, as inflexible and remote as a snow image on a mountain.

'I don't know quite how to make her clear to you, Watson. Perhaps you may meet her before we are through, and you can use your own gift of words. She is beautiful, but with the ethereal other-world beauty of some fanatic whose thoughts are set on high. I have seen such faces in the pictures of the old masters of the Middle Ages. How a beast-man could have laid his vile paws upon such a being of the beyond I cannot imagine. You may have noticed how extremes call to each other, the spiritual to the animal, the cave-man to the angel. You never saw a worse case than this.

'She knew what we had come for, of course—that villain had lost no time in poisoning her mind against us. Miss Winter's advent rather amazed her, I think, but she waved us into our respective chairs like a Reverend Abbess receiving two rather leprous mendicants. If your head is inclined to swell, my dear Watson, take a course of Miss Violet de Merville.

' "Well, sir," said she, in a voice like the wind from an iceberg, "your name is familiar to me. You have called, as I understand, to malign my *fiancé*, Baron Gruner. It is only by my father's request that I see you at all, and I warn you in advance that anything you can say could not possibly have the slightest effect upon my mind."

'I was sorry for her, Watson. I thought of her for the moment as I would have thought of a daughter of my own. I am not often eloquent. I use my head, not my heart. But I really did plead with her with all the warmth of words that I could find in my nature. I pictured to her the awful position of the woman who only wakes to a man's character after she is his wife—a woman who has to submit to be caressed by bloody hands and lecherous lips. I spared her nothing—the shame, the fear, the agony, the hopelessness of it all. All my hot words could not bring one tinge of colour to those ivory cheeks or one gleam of emotion to those abstracted eyes. I thought of what the rascal had said about a post-hypnotic influence. One could really believe that she was living above the earth in some ecstatic dream. Yet there was nothing indefinite in her replies.

' "I have listened to you with patience, Mr Holmes," said she. "The effect upon my mind is exactly as predicted. I am aware that Adelbert, that my *fiancé*, has had a stormy life in which he has incurred bitter hatreds and most unjust aspersions. You are only the last of a series who have brought their slanders before me. Possibly you mean well, though I learn that you are a paid agent who would have been equally willing to act for the Baron as against him. But in any case I wish you to understand once for all that I love him and that he loves me, and that the opinion of all the world is no more to me than the twitter of those birds outside the window. If his noble nature has ever for an instant fallen, it may be that I have been specially sent to raise it to its true and lofty level. I am not clear", here she turned her eyes upon my companion, "who this young lady may be."

'I was about to answer when the girl broke in like a whirlwind. If ever you saw flame and ice face to face, it was those two women.

' "I'll tell you who I am," she cried, springing out of her chair, her mouth all twisted with passion—"I am his last mistress. I am one of a hundred that he has tempted and used and ruined and thrown into the refuse heap, as he will you also. *Your* refuse heap is more likely to be a grave, and maybe that's the best. I tell you, you foolish woman, if you marry this man he'll be the death of you. It may be a broken heart or it may be a broken neck, but he'll have you one way or the other. It's not out of love for you I'm speaking. I don't care a tinker's curse whether you live or die. It's out of hate for him and to spite him and to get back on him for what he did to me. But it's all the same, and you needn't look at me like that, my fine lady, for you may be lower than I am before you are through with it."

' "I should prefer not to discuss such matters," said Miss de Merville, coldly. "Let me say once for all that I am aware of three passages in my *fiancé*'s life in which he became entangled with designing women, and that I am assured of his hearty repentance for any evil that he may have done."

' "Three passages!" screamed my companion. "You fool! You unutterable fool!"

' "Mr Holmes, I beg that you will bring this interview to an end," said the icy voice. "I have obeyed my father's wish in seeing you, but I am not compelled to listen to the ravings of this person."

'With an oath Miss Winter darted forward, and if I had not caught her wrist she would have clutched this maddening woman by the hair. I dragged her towards the door, and was lucky to get her back into the cab without a public scene, for she was beside herself with rage. In a cold way I felt pretty furious myself, Watson, for there was something indescribably annoying in the calm aloofness and supreme self-complaisance of the woman whom we were trying to save. So now once again you know exactly how we stand, and it is clear that I must plan some fresh opening move, for this gambit won't work. I'll keep in touch with you, Watson, for it is more than likely that you will have your part to play, though it is just possible that the next move may lie with them rather than with us.'

And it did. Their blow fell—or his blow rather, for never could I believe that the lady was privy to it. I think I could show you the very paving-stone upon which I stood when my eyes fell upon the placard, and a pang of horror passed through my very soul. It was between the Grand Hotel* and Charing Cross Station,* where a one-legged news-vendor displayed his evening papers. The date was just two days after the last conversation. There, black upon yellow, was the terrible news-sheet:

MURDEROUS
ATTACK
UPON
SHERLOCK
HOLMES

I think* I stood stunned for some moments. Then I have a confused recollection of snatching at a paper, of the remonstrance of the man, whom I had not paid, and, finally, of standing in the doorway of a chemist's shop while I turned up the fateful paragraph. This was how it ran:

> We learn with regret that Mr Sherlock Holmes, the well-known private detective, was the victim this morning of a murderous assault which has left him in a precarious position. There are no exact details to hand, but the event seems to have occurred about twelve o'clock in Regent Street, outside the Café Royal.* The attack was made by two men armed with sticks, and Mr Holmes was beaten about the head and body, receiving injuries which the doctors describe as most serious. He was carried to Charing Cross Hospital,* and afterwards insisted upon being taken to his rooms in Baker Street. The miscreants who attacked him appear to have been respectably dressed men, who escaped from the bystanders by passing through the Café

Royal and out into Glasshouse Street behind it. No doubt they belonged to that criminal fraternity which has so often had occasion to bewail the activity and ingenuity of the injured man.

I need not say that my eyes had hardly glanced over the paragraph before I had sprung into a hansom* and was on my way to Baker Street. I found Sir Leslie Oakshott, the famous surgeon, in the hall and his brougham* waiting at the kerb.

'No immediate danger,' was his report. 'Two lacerated scalp wounds and some considerable bruises. Several stitches have been necessary. Morphine* has been injected and quiet is essential, but an interview of a few minutes would not be absolutely forbidden.'

With this permission I stole into the darkened room. The sufferer was wide awake, and I heard my name in a hoarse whisper. The blind was three-quarters down, but one ray of sunlight slanted through and struck the bandaged head of the injured man. A crimson patch had soaked through the white linen compress.* I sat beside him and bent my head.

'All right, Watson. Don't look so scared,' he muttered in a very weak voice. 'It's not as bad as it seems.'

'Thank God for that!'

'I'm a bit of a single-stick* expert, as you know. I took most of them on my guard. It was the second man that was too much for me.'

'What can I do, Holmes? Of course, it was that damned fellow who set them on. I'll go and thrash the hide off him if you give the word.'

'Good old Watson! No, we can do nothing there unless the police lay their hands on the men. But their get-away had been well prepared. We may be sure of that. Wait a little. I have my plans. The first thing is to exaggerate my injuries. They'll come to you for news. Put it on thick, Watson. Lucky if I live the week out—concussion—delirium—what you like! You can't overdo it.'

'But Sir Leslie Oakshott?'

'Oh, he's all right. He shall see the worst side of me. I'll look after that.'

'Anything else?'

'Yes. Tell Shinwell Johnson to get that girl out of the way. Those beauties will be after her now. They know, of course, that she was with me in the case. If they dared to do me in it is not likely they will neglect her. That is urgent. Do it to-night.'

'I'll go now. Anything more?'

'Put my pipe on the table—and the tobacco-slipper.* Right! Come in each morning and we will plan our campaign.'

I arranged with Johnson that evening to take Miss Winter to a quiet suburb and see that she lay low until the danger was past.

For six days the public were under the impression that Holmes was at the door of death. The bulletins were very grave and there were sinister paragraphs in the papers. My continual visits assured me that it was not so bad as that. His wiry constitution and his determined will were working wonders. He was recovering fast, and I had suspicions at times that he was really finding himself faster than he pretended, even to me. There was a curious secretive streak in the man which led to many dramatic effects, but left even his closest friend guessing as to what his exact plans might be. He pushed to an extreme the axiom that the only safe plotter was he who plotted alone. I was nearer him than anyone else, and yet I was always conscious of the gap between.

On the seventh day the stitches were taken out, in spite of which there was a report of erysipelas* in the evening papers. The same evening papers had an announcement which I was bound, sick or well, to carry to my friend. It was simply that among the passengers on the Cunard boat* *Ruritania*,* starting from Liverpool on Friday, was the Baron Adelbert Gruner, who had some important financial business to settle in the States before his impending wedding to Miss Violet de Merville, only daughter of, etc., etc. Holmes listened to the news with a cold, concentrated look upon his pale face, which told me that it hit him hard.

'Friday!' he cried. 'Only three clear days. I believe the rascal wants to put himself out of danger's way. But he won't, Watson! By the Lord Harry, he won't! Now, Watson, I want you to do something for me.'

'I am here to be used, Holmes.'

'Well, then, spend the next twenty-four hours in an intensive study of Chinese pottery.'

He gave no explanations and I asked for none. By long experience I had learned the wisdom of obedience. But when I had left his room I walked down Baker Street, revolving in my head how on earth I was to carry out so strange an order. Finally I drove to the London Library* in St James's Square,* put the matter to my friend Lomax, the sub-librarian, and departed to my rooms with a goodly volume* under my arm.

It is said that the barrister who crams up a case with such care that he can examine an expert witness upon the Monday has forgotten all his forced knowledge before the Saturday. Certainly I should not like now to pose as an authority upon ceramics. And yet all that evening, and all that night with a short interval for rest, and all next morning I was sucking in knowledge and committing names to memory. There I learned of the hall-marks of the great artist-decorators,* of the mystery of cyclical dates, the marks of the Hung-wu and the beauties of the Yung-lo, the writings of Tang-ying, and the glories of the primitive period of the Sung and the Yuan.* I was charged with all this information when I called upon Holmes next evening. He was out of bed now, though you would not have guessed it from the published reports, and he sat with his much-bandaged head resting upon his hand in the depth of his favourite arm-chair.

'Why, Holmes,' I said, 'if one believed the papers you are dying.'

'That', said he, 'is the very impression which I intended to convey. And now, Watson, have you learned your lessons?'

'At least I have tried to.'

'Good. You could keep up an intelligent conversation on the subject?'

'I believe I could.'

'Then hand me that little box from the mantelpiece.'

He opened the lid and took out a small object most carefully wrapped in some fine Eastern silk. This he unfolded, and disclosed a delicate little saucer of the most beautiful deep-blue colour.

'It needs careful handling, Watson. This is the real eggshell pottery* of the Ming dynasty. No finer piece ever passed through Christie's. A complete set of this would be worth a king's ransom—in fact, it is doubtful if there is a complete set outside the Imperial palace of Peking. The sight of this would drive a real connoisseur wild.'

'What am I to do with it?'

Holmes handed me a card upon which was printed: 'Dr Hill Barton,* 369, Half Moon Street.'

'That is your name for the evening, Watson. You will call upon Baron Gruner. I know something of his habits, and at half-past eight he would probably be disengaged. A note will tell him in advance that you are about to call, and you will say that you are bringing him a specimen of an absolutely unique set of Ming china. You may as well be a medical man, since that is a part which you can play without duplicity. You are a collector, this set has come your way, you have heard of the Baron's interest in the subject, and you are not averse to selling at a price.'

'What price?'

'Well asked, Watson. You would certainly fall down badly if you did not know the value of your own wares. This saucer was got for me by Sir James, and comes, I understand, from the collection of his client. You will not exaggerate if you say that it could hardly be matched in the world.'

'I could perhaps suggest that the set should be valued by an expert.'

'Excellent, Watson! You scintillate today. Suggest Christie or Sotheby.* Your delicacy prevents your putting a price for yourself.'

'But if he won't see me?'

'Oh, yes, he will see you. He has the collection mania in its most acute form—and especially on this subject, on which he is an acknowledged authority. Sit down, Watson, and I will dictate the letter. No answer needed. You will merely say that you are coming, and why.'

It was an admirable document, short, courteous, and stimulating to the curiosity of the connoisseur. A district messenger was duly dispatched with it. On the same evening, with the precious saucer in my hand and the card of Dr Hill Barton in my pocket, I set off on my own adventure.

The beautiful house and grounds indicated that Baron Gruner was, as Sir James had said, a man of considerable wealth. A long winding drive, with banks of rare shrubs on either side, opened out into a great gravelled square adorned with statues. The place had been built by a South African gold king* in the days of the great boom,* and the long, low house with the turrets at the corners, though an architectural nightmare, was imposing in its size and solidity. A butler who would have adorned a bench of Bishops showed me in, and handed me over to a plush-clad footman,* who ushered me into the Baron's presence.

He was standing at the open front of a great case which stood between the windows, and which contained part of his Chinese collection. He turned as I entered with a small brown vase* in his hand.

'Pray sit down, doctor,' said he. 'I was looking over my own treasures and wondering whether I could really afford to add to them. This little Tang specimen, which dates from the seventh century, would probably interest you. I am sure you never saw finer workmanship or a richer glaze. Have you the Ming saucer with you of which you spoke?'

I carefully unpacked it and handed it to him. He seated himself at his desk, pulled over the lamp, for it was growing dark, and set himself to examine it. As he did so the yellow light beat upon his own features, and I was able to study them at my ease.

He was certainly a remarkably handsome man. His European reputation for beauty was fully deserved. In figure he was not more than of middle size, but was built upon graceful and active lines. His face was swarthy, almost Oriental, with large, dark, languorous eyes which might easily hold an irresistible fascination for women. His hair and moustache were raven black, the latter short, pointed, and carefully waxed. His features were regular and pleasing, save only his straight, thin-lipped mouth. If ever I saw a murderer's mouth it was there—a cruel, hard gash in the face, compressed, inexorable, and terrible. He was ill-advised to train his moustache away from it, for it was Nature's danger-signal, set as a warning to his victims. His voice was engaging and his manners perfect. In age I should have put him at little over thirty, though his record afterwards showed that he was forty-two.

'Very fine—very fine indeed!' he said at last. 'And you say you have a set of six to correspond. What puzzles me is that I should not have heard of such magnificent specimens. I only know of one in England to match this, and it is certainly not likely to be in the market. Would it be indiscreet if I were to ask you, Dr Hill Barton, how you obtained this?'

'Does it really matter?' I asked, with as careless an air as I could muster. 'You can see that the piece is genuine, and, as to the value, I am content to take an expert's valuation.'

'Very mysterious,' said he, with a quick suspicious flash, of his dark eyes. 'In dealing with objects of such value, one naturally wishes to know all about the transaction. That the piece is genuine is certain. I have no doubts at all about that. But suppose—I am bound to take every possibility into account—that it should prove afterwards that you had no right to sell?'

'I would guarantee you against any claim of the sort.'

'That, of course, would open up the question as to what your guarantee was worth.'

'My bankers would answer that.'

'Quite so. And yet the whole transaction strikes me as rather unusual.'

'You can do business or not,' said I, with indifference. 'I have given you the first offer as I understood that you were a connoisseur, but I shall have no difficulty in other quarters.'

'Who told you I was a connoisseur?'

'I was aware that you had written a book upon the subject.'

'Have you read the book?'

'No.'

'Dear me, this becomes more and more difficult for me to understand! You are a connoisseur and collector with a very valuable piece in your collection, and yet you have never troubled to consult the one book which would have told you of the real meaning and value of what you held. How do you explain that?'

'I am a very busy man. I am a doctor in practice.'

'That is no answer. If a man has a hobby he follows it up, whatever his other pursuits may be. You said in your note that you were a connoisseur.'

'So I am.'

'Might I ask you a few questions to test you? I am obliged to tell you, doctor—if you are indeed a doctor—that the incident becomes more and more suspicious. I would ask you what do you know of the Emperor Shomu and how do you associate him with the Shoso-in near Nara?* Dear me, does that puzzle you? Tell me a little about the Northern Wei dynasty* and its place in the history of ceramics.'

I sprang from my chair in simulated anger.

'This is intolerable, sir,' said I. 'I came here to do you a favour, and not to be examined as if I were a schoolboy. My knowledge on these subjects may be second only to your own, but I certainly shall not answer questions which have been put in so offensive a way.'

He looked at me steadily. The languor had gone from his eyes. They suddenly glared. There was a gleam of teeth from between those cruel lips.

'What is the game? You are here as a spy. You are an emissary of Holmes. This is a trick that you are playing upon me. The fellow is dying, I hear, so he sends his tools to keep watch upon me. You've made your way in here without leave, and by God! you may find it harder to get out than to get in.'

He had sprung to his feet, and I stepped back, bracing myself for an attack, for the man was beside himself with rage. He may have suspected me from the first; certainly this cross-examination had shown him the truth; but it was clear that I could not hope to deceive him. He dived his hand into a side-drawer and rummaged furiously. Then something struck upon his ear, for he stood listening intently.

'Ah!' he cried. 'Ah!' and dashed into the room behind him.

Two steps took me to the open door, and my mind will ever carry a clear picture of the scene within. The window leading out to the garden was wide open. Beside it, looking like some terrible ghost, his head girt with bloody bandages, his face drawn and white, stood Sherlock Holmes. The next instant he was through the gap, and I heard the crash of his body among the laurel bushes outside. With a howl of rage the master of the house rushed after him to the open window.

And then! It was done in an instant, and yet I clearly saw it. An arm—a woman's arm—shot out from among the leaves. At the same instant the Baron uttered a horrible cry—a yell which will always ring in my memory. He clapped his two hands to his face and rushed round the room, beating his head horribly against the walls. Then he fell upon the carpet, rolling and writhing, while scream after scream resounded through the house.

'Water! For God's sake, water!' was his cry.

I seized a carafe from a side-table and rushed to his aid. At the same moment the butler and several footmen ran in from the hall. I remember that one of them fainted as I knelt by the injured man and turned that awful face to the light of the lamp. The vitriol* was eating into it everywhere and dripping from the ears and the chin. One eye was already white and glazed. The other was red and inflamed. The features which I had admired a few minutes before were now like some beautiful painting over which the artist has passed a wet and foul sponge. They were blurred, discoloured, inhuman, terrible.

In a few words I explained exactly what had occurred, so far as the vitriol attack was concerned. Some had climbed through the window and others had rushed out on to the lawn, but it was dark and it had begun to rain. Between his screams the victim raged and raved against the avenger. 'It was that hell-cat, Kitty Winter!' he cried. 'Oh, the she-devil! She shall pay for it! She shall pay! Oh, God in heaven, this pain is more than I can bear!'

I bathed his face in oil, put cotton wadding on the raw surfaces, and administered a hypodermic of morphia.* All suspicion of me had passed from his mind in the presence of this shock, and he clung to my hands as if I might have the power even yet to clear those dead-fish eyes which gazed up at me. I could have wept over the ruin had I not remembered very clearly the vile life which had led up to so hideous a change. It was loathsome to feel the pawing of his burning hands, and I was relieved when his family surgeon, closely followed by a specialist, came to relieve me of my charge. An inspector of police had also arrived, and to him I handed my real card. It would have been useless as well as foolish to do otherwise, for I was nearly as well known by sight at the Yard as Holmes himself. Then I left that house of gloom and terror. Within an hour I was at Baker Street.

Holmes was seated in his familiar chair, looking very pale and exhausted. Apart from his injuries, even his iron nerves had been shocked by the events of the evening, and he listened with horror to my account of the Baron's transformation.

'The wages of sin,* Watson—the wages of sin!' said he. 'Sooner or later it will always come. God knows, there was sin enough,' he added, taking up a brown volume from the table. 'Here is the book the woman talked of. If this will not break off the marriage, nothing ever could. But it will, Watson. It must. No self-respecting woman could stand it.'

'It is his love diary?'

'Or his lust diary. Call it what you will. The moment the woman told us of it I realized what a tremendous weapon was there, if we could but lay our hands on it. I said nothing at the time to indicate my thoughts, for this woman might have given it away. But I brooded over it. Then this assault upon me gave me the chance of letting the Baron think that no precautions need be taken against me. That was all to the good. I would have waited a little longer, but his visit to America forced my hand. He would never have left so compromising a document behind him. Therefore we had to act at once. Burglary at night is impossible. He takes precautions. But there was a chance in the evening if I could only be sure that his attention was engaged. That was where you and your blue saucer came in. But I had to be sure of the position of the book, and I knew I had only a few minutes in which to act, for my time was limited by your knowledge of Chinese pottery. Therefore I gathered the girl up at the last moment. How could I guess what the little packet was that she carried so carefully under her cloak? I thought she had come altogether on my business, but it seems she had some of her own.'

'He guessed I came from you.'

'I feared he would. But you held him in play just long enough for me to get the book, though not long enough for an unobserved escape. Ah, Sir James, I am very glad you have come!'

Our courtly friend had appeared in answer to a previous summons. He listened with the deepest attention to Holmes's account of what had occurred.

'You have done wonders—wonders!' he cried, when he had heard the narrative. 'But if these injuries are as terrible as Dr Watson describes, then surely our purpose of thwarting the marriage is sufficiently gained without the use of this horrible book.'

Holmes shook his head.

'Women of the de Merville type do not act like that. She would love him the more as a disfigured martyr. No, no. It is his moral side, not his physical, which we have to destroy. That book will bring her back to earth—and I know nothing else that could. It is in his own writing. She cannot get past it.'

Sir James carried away both it and the precious saucer. As I was myself overdue,* I went down with him into the street. A brougham was waiting for him. He sprang in, gave a hurried order to the cockaded coachman,* and drove swiftly away. He flung his overcoat half out of the window to cover the armorial bearings* upon the panel, but I had seen them in the glare of our fanlight none the less. I gasped with surprise. Then I turned back and ascended the stair to Holmes's room.

'I have found out who our client is,' I cried, bursting with my great news. 'Why, Holmes, it is—'

'It is a loyal friend and a chivalrous gentleman,' said Holmes, holding up a restraining hand. 'Let that now and for ever be enough for us.'

I do not know how the incriminating book was used. Sir James may have managed it. Or it is more probable that so delicate a task was entrusted to the young lady's father. The effect, at any rate, was all that could be desired. Three days later appeared a paragraph in *The Morning Post* to say that the marriage between Baron Adelbert Gruner and Miss Violet de Merville would not take place. The same paper had the first police-court hearing of the proceedings against Miss Kitty Winter on the grave charge of vitriol-throwing. Such extenuating circumstances came out in the trial that the sentence, as will be remembered, was the lowest that was possible for such an offence. Sherlock Holmes was threatened with a prosecution for burglary, but when an object is good and a client is sufficiently illustrious, even the rigid British law becomes human and elastic. My friend has not yet stood in the dock.

The Three Gables

I DON'T think that any of my adventures with Mr Sherlock Holmes opened quite so abruptly, or so dramatically, as that which I associate with The Three Gables.* I had not seen Holmes for some days, and had no idea of the new channel into which his activities had been directed. He was in a chatty mood that morning, however, and had just settled me into the well-worn low arm-chair on one side of the fire, while he had curled down with his pipe in his mouth upon the opposite chair, when our visitor arrived. If I had said that a mad bull had arrived, it would give a clearer impression of what occurred.

The door had flown open and a huge negro* had burst into the room. He would have been a comic figure if he had not been terrific, for he was dressed in a very loud grey check suit with a flowing salmon-coloured tie. His broad face and flattened nose were thrust forward, as his sullen dark eyes, with a smouldering gleam of malice in them, turned from one of us to the other.

'Which of you genelmen is Masser Holmes?' he asked.

Holmes raised his pipe with a languid smile.

'Oh! it's you, is it?' said our visitor, coming with an unpleasant, stealthy step round the angle of the table. 'See here, Masser Holmes, you keep your hands out of other folks' business. Leave folks to manage their own affairs. Got that, Masser Holmes?'*

'Keep on talking,' said Holmes. 'It's fine.'

'Oh! it's fine, is it?' growled the savage. 'It won't be so damn fine if I have to trim you up a bit. I've handled your kind before now, and they didn't look fine when I was through with them. Look at that, Masser* Holmes!'

He swung a huge knotted lump of a fist under my friend's nose. Holmes examined it closely with an air of great interest.

'Were you born so?' he asked. 'Or did it come by degrees?'

It may have been the icy coolness of my friend, or it may have been the slight clatter which I made as I picked up the poker. In any case, our visitor's manner became less flamboyant.

'Well, I've given you fair warnin',' said he. 'I've a friend that's interested out Harrow way—you know what I'm meaning—and he don't intend to have no buttin' in by you. Got that? You ain't the law, and I ain't the law either, and if you come in I'll be on hand also. Don't you forget it.'

'I've wanted to meet you for some time,' said Holmes. 'I won't ask you to sit down, for I don't like the smell of you, but aren't you Steve Dixie, the bruiser?'

'That's my name, Masser Holmes, and you'll get put through it for sure if you give me any lip.'

'It is certainly the last thing you need,' said Holmes, staring at our visitor's hideous mouth. 'But it was the killing of young Perkins outside the Holborn Bar*—What! you're not going?'

The negro had sprung back, and his face was leaden. 'I won't listen to no such talk,' said he. 'What have I to do with this 'ere Perkins, Masser Holmes? I was trainin' at the Bull Ring in* Birmingham when this boy done gone get into trouble.'

'Yes, you'll tell the magistrate about it, Steve,' said Holmes. 'I've been watching you and Barney Stockdale—'

'So help me the Lord! Masser Holmes—'

'That's enough. Get out of it. I'll pick you up when I want you.'

'Good mornin', Masser Holmes. I hope there ain't no hard feelin's about this 'ere visit?'

'There will be unless you tell me who sent you.'

'Why, there ain't no secret about that, Masser Holmes. It was the same gentleman that you have just done gone mention.'

'And who set him on to it?'

'S'elp me. I don't know, Masser Holmes. He just say, "Steve, you go see Mr Holmes, and tell him his life ain't safe if he go down Harrow way." That's the whole truth.'

Without waiting for any further questioning, our visitor bolted out of the room almost as precipitately as he had entered. Holmes knocked out the ashes of his pipe with a quiet chuckle.

'I am glad you were not forced to break his woolly head, Watson. I observed your manoeuvres with the poker. But he is really rather a harmless fellow, a great muscular, foolish, blustering baby, and easily cowed, as you have seen. He is one of the Spencer John* gang and has taken part in some dirty work of late which I may clear up when I have time. His immediate principal, Barney, is a more astute person. They specialize in assaults, intimidation, and the like. What I want to know is, who is at the back of them on this particular occasion?'

'But why do they want to intimidate you?'

'It is this Harrow Weald case. It decides me to look into the matter, for if it is worth anyone's while to take so much trouble, there must be something in it.'

'But what is it?'

'I was going to tell you when we had this comic interlude. Here is Mrs Maberley's note. If you care to come with me we will wire her and go out at once.'

DEAR MR SHERLOCK HOLMES, [I read]

I have had a succession of strange incidents occur to me in connection with this house, and I should much value your advice. You would find me at home any time to-morrow. The house is within a short walk of the Weald Station. I believe that my late husband, Mortimer Maberley, was one of your early clients.

Yours faithfully,
MARY MABERLEY

The address was 'The Three Gables, Harrow Weald'.*

'So that's that!' said Holmes. 'And now if you can spare the time, Watson, we will get upon our way.'

A short railway journey, and a shorter drive, brought us to the house, a brick and timber villa, standing in its own acre of undeveloped grassland. Three small projections above the upper windows made a feeble attempt to justify its name. Behind was a grove of melancholy, half-grown pines, and the whole aspect of the place was poor and depressing. None the less, we found the house to be well furnished, and the lady who received us was a most engaging elderly person, who bore every mark of refinement and culture.

'I remember your husband well, madam,' said Holmes, 'though it is some years since he used my services in some trifling matter.'

'Probably you would be more familiar with the name of my son Douglas.'

Holmes looked at her with great interest.

'Dear me! Are you the mother of Douglas Maberley? I knew him slightly. But, of course, all London knew him. What a magnificent creature he was! Where is he now?'

'Dead, Mr Holmes, dead! He was Attaché* at Rome, and he died there of pneumonia last month.'

'I am sorry. One could not connect death with such a man. I have never known anyone so vitally alive. He lived intensely—every fibre of him!'

'Too intensely, Mr Holmes. That was the ruin of him. You remember him as he was—debonair and splendid. You did not see the moody, morose, brooding creature into which he developed. His heart was broken. In a single month I seemed to see my gallant boy turn into a worn-out cynical man.'

'A love affair—a woman?'

'Or a fiend. Well, it was not to talk of my poor lad that I asked you to come, Mr Holmes.'

'Dr Watson and I are at your service.'

'There have been some very strange happenings. I have been in this house more than a year now, and as I wished to lead a retired life I have seen little of my neighbours. Three days ago I had a call from a man who said that he was a house agent. He said that this house would exactly suit a client of his and that if I would part with it money would be no object. It seemed to me very strange, as there are several empty houses on the market which appear to be equally eligible, but naturally I was interested in what he said. I therefore named a price which was five hundred pounds more than I gave. He at once closed with the offer, but added that his client desired to buy the furniture as well and would I put a price upon it. Some of this furniture is from my old home, and it is, as you see, very good, so that I named a good round sum. To this also he at once agreed. I had always wanted to travel, and the bargain was so good a one that it really seemed that I should be my own mistress for the rest of my life.

'Yesterday the man arrived with the agreement all drawn out. Luckily I showed it to Mr Sutro, my lawyer, who lives in Harrow. He said to me, "This is a very strange document. Are you aware that if you sign it you could not legally take *anything* out of the house—not even your own private possessions?" When the man came again in the evening I pointed this out, and I said that I meant only to sell the furniture.

' "No, no; everything," said he.

' "But my clothes? My jewels?"

' "Well, well, some concession might be made for your personal effects. But nothing shall go out of the house unchecked. My client is a very liberal man, but he has his fads and his own way of doing things. It is everything or nothing with him."

' "Then it must be nothing," said I. And there the matter was left, but the whole thing seemed to me to be so unusual that I thought—'

Here we had a very extraordinary interruption.

Holmes raised his hand for silence. Then he strode across the room, flung open the door, and dragged in a great gaunt woman whom he had seized by the shoulder. She entered with ungainly struggles, like some huge awkward chicken, torn squawking out of its coop.

'Leave me alone! What are you a-doin' of?' she screeched.

'Why, Susan, what is this?'

'Well, ma'am, I was comin' in to ask if the visitors was stayin' for lunch when this man jumped out at me.'

'I have been listening to her for the last five minutes, but did not wish to interrupt your most interesting narrative.

Just a little wheezy, Susan, are you not? You breathe too heavily for that kind of work.'

Susan turned a sulky but amazed face upon her captor. 'Who be you, anyhow, and what right have you a-pullin' me about like this?'

'It was merely that I wished to ask a question in your presence. Did you, Mrs Maberley, mention to anyone that you were going to write to me and consult me?'

'No. Mr Holmes, I did not.'

'Who posted your letter?'

'Susan did.'

'Exactly. Now, Susan, to whom was it that you wrote or sent a message to say that your mistress was asking advice from me?'

'It's a lie. I sent no message.'

'Now, Susan, wheezy people may not live long, you know. It's a wicked thing to tell fibs. Whom did you tell?'

'Susan!' cried her mistress, 'I believe you are a bad, treacherous woman. I remember now that I saw you speaking to someone over the hedge.'

'That was my own business,' said the woman sullenly.

'Suppose I tell you that it was Barney Stockdale to whom you spoke?' said Holmes.

'Well, if you know, what do you want to ask for?'

'I was not sure, but I know now. Well now, Susan, it will be worth ten pounds to you if you will tell me who is at the back of Barney.'

'Someone that could lay down a thousand pounds for every ten you have in the world.'

'So, a rich man? No; you smiled—a rich woman. Now we have got so far, you may as well give the name and earn the tenner.'

'I'll see you in hell first.'

'Oh, Susan! Language!'

'I am clearing out of here. I've had enough of you all. I'll send for my box to-morrow.' She flounced for the door.

'Good-bye, Susan. Paregoric* is the stuff . . . Now,' he continued, turning suddenly from lively to severe when the door had closed behind the flushed and angry woman, 'this gang means business. Look how close they play the game. Your letter to me had the 10 p.m. postmark. And yet Susan passes the word to Barney. Barney has time to go to his employer and get instructions; he or she—I incline to the latter from Susan's grin when she thought I had blundered—forms a plan. Black Steve is called in, and I am warned off by eleven o'clock next morning. That's quick work, you know.'

'But what do they want?'

'Yes, that's the question. Who had the house before you?'

'A retired sea captain, called Ferguson.'

'Anything remarkable about him?'

'Not that ever I heard of.'

'I was wondering whether he could have buried something. Of course, when people bury treasure nowadays they do it in the Post Office bank. But there are always some lunatics about. It would be a dull world without them. At first I thought of some buried valuable. But why, in that case, should they want your furniture? You don't happen to have a Raphael* or a first folio Shakespeare* without knowing it?'

'No, I don't think I have anything rarer than a Crown Derby tea-set.'*

'That would hardly justify all this mystery. Besides, why should they not openly state what they want? If they covet your tea-set, they can surely offer a price for it without buying you out, lock, stock, and barrel.* No, as I read it, there is something which you do not know that you have, and which you would not give up if you did know.'

'That is how I read it,' said I.

'Dr Watson agrees, so that settles it.'

'Well, Mr Holmes, what can it be?'

'Let us see whether by this purely mental analysis we can get it to a finer point. You have been in this house a year.'

'Nearly two.'

'All the better. During this long period no one wants anything from you. Now suddenly within three or four days you have urgent demands. What would you gather from that?'

'It can only mean,' said I, 'that the object, whatever it may be, has only just come into the house.'

'Settled once again,' said Holmes. 'Now, Mrs Maberley, has any object just arrived?'

'No; I have bought nothing new this year.'

'Indeed! That is very remarkable. Well, I think we had best let matters develop a little further until we have clearer data. Is that lawyer of yours a capable man?'

'Mr Sutro is most capable.'

'Have you another maid, or was the fair Susan, who has just banged your front door, alone?'

'I have a young girl.'

'Try and get Sutro to spend a night or two in the house. You might possibly want protection.'

'Against whom?'

'Who knows? The matter is certainly obscure. If I can't find what they are after, I must approach the matter from the other end, and try to get at the principal. Did this house-agent man give any address?'

'Simply his card and occupation. Haines-Johnson, Auctioneer and Valuer.'*

'I don't think we shall find him in the Directory. Honest businessmen don't conceal their place of business. Well, you will let me know any fresh development. I have taken up your case, and you may rely upon it that I shall see it through.'

As we passed through the hall Holmes's eyes, which missed nothing, lighted upon several trunks and cases which were piled in the corner. The labels shone out upon them.

' "Milano." "Lucerne."* These are from Italy.'

'They are poor Douglas's things.'

'You have not unpacked them? How long have you had them?'

'They arrived last week.'

'But you said—why, surely this might be the missing link. How do we know that there is not something of value there?'

'There could not possibly be, Mr Holmes. Poor Douglas had only his pay and a small annuity. What could he have of value?'

Holmes was lost in thought.

'Delay no longer, Mrs Maberley,' he said at last. 'Have these things taken upstairs to your bedroom. Examine them as soon as possible and see what they contain. I will come to-morrow and hear your report.'

It was quite evident that The Three Gables was under very close surveillance, for as we came round the high hedge at the end of the lane there was the negro prize-fighter standing in the shadow. We came on him quite suddenly, and a grim and menacing figure he looked in that lonely place. Holmes clapped his hand to his pocket.

'Lookin' for your gun, Masser Holmes?'

'No; for my scent-bottle, Steve.'

'You are funny, Masser Holmes, ain't you?'

'It won't be funny for you, Steve, if I get after you. I gave you fair warning this morning.'

'Well, Masser Holmes, I done gone think over what you said, and I don't want no more talk about that affair of Masser Perkins. S'pose I can help you, Masser Holmes, I will.'

'Well, then, tell me who is behind you on this job?'

'So help me the Lord! Masser Holmes, I told you the truth before. I don't know. My boss Barney gives me orders and that's all.'

'Well, just bear in mind, Steve, that the lady in that house, and everything under that roof, is under my protection. Don't you forget it.'

'All right, Masser Holmes. I'll remember.'

'I've got him thoroughly frightened for his own skin, Watson,' Holmes remarked as we walked on. 'I think he would double-cross his employer if he knew who he was. It was lucky I had some knowledge of the Spencer John crowd, and that Steve was one of them. Now, Watson, this is a case for Langdale Pike,* and I am going to see him now. When I get back I may be clearer in the matter.'

I saw no more of Holmes during the day, but I could well imagine how he spent it, for Langdale Pike was his human book of reference upon all matters of social scandal. This strange, languid creature spent his waking hours in the bow window of a St James's Street club, and was the receiving-station, as well as the transmitter, for all the gossip of the Metropolis. He made, it was said, a four-figure income by the paragraphs which he contributed every week to the garbage papers which cater for an inquisitive public. If ever, far down in the turbid depths of London life, there was some strange swirl or eddy, it was marked with automatic exactness by this human dial upon the surface. Holmes discreetly helped Langdale to knowledge, and on occasion was helped in turn.

When I met my friend in his room early next morning, I was conscious from his bearing that all was well, but none the less a most unpleasant surprise was awaiting us. It took the shape of the following telegram:

> Please come out at once. Client's house burgled in the night. Police in possession.
>
> SUTRO

Holmes whistled. 'The drama has come to a crisis, and quicker than I had expected. There is a great driving-power at the back of this business, Watson, which does not surprise me after what I have heard. This Sutro, of course, is her lawyer. I made a mistake, I fear, in not asking you to spend the night on guard. This fellow has clearly proved a broken reed.* Well, there is nothing for it but another journey to Harrow Weald.'

We found The Three Gables a very different establishment to the orderly household of the previous day. A small group of idlers had assembled at the garden gate, while a couple of constables were examining the windows and the geranium beds. Within we met a grey old gentleman, who introduced himself as the lawyer, together with a bustling, rubicund Inspector, who greeted Holmes as an old friend.

'Well, Mr Holmes, no chance for you in this case, I'm afraid. Just a common, ordinary burglary, and well within the capacity of the poor old police. No experts need apply.'

'I am sure the case is in very good hands,' said Holmes. 'Merely a common burglary, you say?'

'Quite so. We know pretty well who the men are and where to find them. It is that gang of Barney Stockdale, with the big nigger in it—they've been seen about here.'

'Excellent! What did they get?'

'Well, they don't seem to have got much. Mrs Maberley was chloroformed* and the house was—Ah! here is the lady herself.'

Our friend of yesterday, looking very pale and ill, had entered the room, leaning upon a little maid-servant.

'You gave me good advice, Mr Holmes,' said she, smiling ruefully. 'Alas, I did not take it! I did not wish to trouble Mr Sutro, and so I was unprotected.'

'I only heard of it this morning,' the lawyer explained.

'Mr Holmes advised me to have some friend in the house. I neglected his advice, and I have paid for it.'

'You look wretchedly ill,' said Holmes. 'Perhaps you are hardly equal to telling me what occurred.'

'It is all here,' said the Inspector, tapping a bulky note-book.

'Still, if the lady is not too exhausted—'

'There is really so little to tell. I have no doubt that wicked Susan had planned an entrance for them. They must have known the house to an inch. I was conscious for a moment of the chloroform rag which was thrust over my mouth, but I have no notion how long I may have been senseless. When I woke, one man was at the bedside and another was rising with a bundle in his hand from among my son's baggage, which was partially opened and littered over the floor. Before he could get away I sprang up and seized him.'

'You took a big risk,' said the Inspector.

'I clung to him, but he shook me off, and the other may have struck me, for I can remember no more. Mary the maid heard the noise and began screaming out of the window. That brought the police, but the rascals had got away.'

'What did they take?'

'Well, I don't think there is anything of value missing. I am sure there was nothing in my son's trunks.'

'Did the men leave no clue?'

'There was one sheet of paper which I may have torn from the man that I grasped. It was lying all crumpled on the floor. It is in my son's handwriting.'

'Which means that it is not of much use,' said the Inspector. 'Now if it had been the burglar's—'

'Exactly,' said Holmes. 'What rugged common sense! None the less, I should be curious to see it.'

The Inspector drew a folded sheet of foolscap from his pocket-book.

'I never pass anything, however trifling,' said he, with some pomposity. 'That is my advice to you, Mr Holmes. In twenty-five years' experience I have learned my lesson. There is always the chance of finger-marks or something.'

Holmes inspected the sheet of paper.

'What do you make of it, Inspector?'

'Seems to be the end of some queer novel, so far as I can see.'

'It may certainly prove to be the end of a queer tale,' said Holmes. 'You have noticed the number on the top of the page. It is two hundred and forty-five. Where are the odd two hundred and forty-four pages?'

'Well, I suppose the burglars got those. Much good may it do them!'

'It seems a queer thing to break into a house in order to steal such papers as that. Does it suggest anything to you, Inspector?'

'Yes, sir; it suggests that in their hurry the rascals just grabbed at what came first to hand. I wish them joy of what they got.'

'Why should they go to my son's things?' asked Mrs Maberley.

'Well, they found nothing valuable downstairs, so they tried their luck upstairs. That is how I read it. What do you make of it, Mr Holmes?'

'I must think it over, Inspector. Come to the window, Watson.' Then, as we stood together, he read over the fragment of paper. It began in the middle of a sentence and ran like this:

. . . face bled considerably from the cuts and blows, but it was nothing to the bleeding of his heart as he saw that lovely face, the face for which he had been prepared to sacrifice his very life, looking out at his agony and humiliation. She smiled—yes, by Heaven! she smiled, like the heartless fiend she was, as he looked up at her. It was at that moment that love died and hate was born. Man must live for something. If it is not for your embrace, my lady, then it shall surely be for your undoing and my complete revenge.

'Queer grammar!' said Holmes, with a smile, as he handed the paper back to the Inspector. 'Did you notice how the "he" suddenly changed to "my"? The writer was so carried away by his own story that he imagined himself at the supreme moment to be the hero.'

'It seemed mighty poor stuff,'* said the Inspector, as he replaced it in his book. 'What! are you off, Mr Holmes?'

'I don't think there is anything more for me to do now that the case is in such capable hands. By the way, Mrs Maberley, did you say you wished to travel?'

'It has always been my dream, Mr Holmes.'

'Where would you like to go—Cairo, Madeira, the Riviera?'

'Oh! if I had the money I would go round the world.'

'Quite so. Round the world. Well, good morning. I may drop you a line in the evening.' As we passed the window I caught a glimpse of the Inspector's smile and shake of the head. 'These clever fellows have always a touch of madness.' That was what I read in the Inspector's smile.

'Now, Watson, we are at the last lap of our little journey,' said Holmes, when we were back in the roar of Central London once more. 'I think we had best clear the matter up at once, and it would be well that you should come with me, for it is safer to have a witness when you are dealing with such a lady as Isadora Klein.'

We had taken a cab and were speeding to some address in Grosvenor Square.* Holmes had been sunk in thought, but he roused himself suddenly.

'By the way, Watson, I suppose you see it all clearly?'

'No, I can't say that I do. I only gather that we are going to see the lady who is behind all this mischief.'

'Exactly! But does the name Isadora Klein convey nothing to you? She was, of course, *the* celebrated beauty. There was never a woman to touch her. She is pure Spanish, the real blood of the masterful conquistadores,* and her people have been leaders in Pernambuco* for generations. She married the aged German sugar king, Klein, and presently found herself the richest as well as the most lovely widow upon earth. Then there was an interval of adventure when she pleased her own tastes. She had several lovers, and Douglas Maberley, one of the most striking men in London, was one of them. It was by all accounts more than an adventure with him. He was not a Society butterfly, but a strong, proud man who gave and expected all. But she is the "*belle dame sans merci*"* of fiction. When her caprice is satisfied, the matter is ended, and if the other party in the matter can't take her word for it, she knows how to bring it home to him.'

'Then that was his own story—'

'Ah! you are piecing it together now. I hear that she is about to marry the young Duke of Lomond,* who might almost be her son. His Grace's ma might overlook the age, but a big scandal would be a different matter, so it is imperative—Ah! here we are.'

It was one of the finest corner-houses of the West End. A machine-like footman took up our cards and returned with word that the lady was not at home. 'Then we shall wait until she is,' said Holmes cheerfully.

The machine broke down.

'Not at home means not at home to *you*,' said the footman.

'Good,' Holmes answered. 'That means that we shall not have to wait. Kindly give this note to your mistress.'

He scribbled three or four words upon a sheet of his notebook, folded it, and handed it to the man.

'What did you say, Holmes?' I asked.

'I simply wrote "Shall it be the police, then?" I think that should pass us in.'

It did—with amazing celerity. A minute later we were in an Arabian Nights* drawing-room, vast and wonderful, in a half gloom, picked out with an occasional pink electric light. The lady had come, I felt, to that time of life when even the proudest beauty finds the half-light more welcome. She rose from the settee as we entered: tall, queenly, a perfect figure, a lovely mask-like face, with two wonderful Spanish eyes which looked murder at us both.

'What is this intrusion—and this insulting message?' she asked, holding up the slip of paper.

'I need not explain, madame. I have too much respect for your intelligence to do so—though I confess that intelligence has been surprisingly at fault of late.'

'How so, sir?'

'By supposing that your hired bullies could frighten me from my work. Surely no man would take up my profession if it were not that danger attracts him. It was you, then, who forced me to examine the case of young Maberley.'

'I have no idea what you are talking about. What have I to do with hired bullies?'

Holmes turned away wearily.

'Yes, I have overrated* your intelligence. Well, good afternoon!'

'Stop! Where are you going?'

'To Scotland Yard.'

We had not got half-way to the door before she had overtaken us and was holding his arm. She had turned in a moment from steel to velvet.

'Come and sit down, gentlemen. Let us talk this matter over. I feel that I may be frank with you, Mr Holmes. You have the feelings of a gentleman. How quick a woman's instinct is to find it out. I will treat you as a friend.'

'I cannot promise to reciprocate, madame. I am not the law, but I represent justice so far as my feeble powers go. I am ready to listen, and then I will tell you how I will act.'

'No doubt it was foolish of me to threaten a brave man like yourself.'

'What was really foolish, madame, is that you have placed yourself in the power of a band of rascals who may blackmail or give you away.'

'No, no! I am not so simple. Since I have promised to be frank, I may say that no one, save Barney Stockdale and Susan, his wife, have the least idea who their employer is. As to them, well, it is not the first—' She smiled and nodded, with a charming coquettish intimacy.

'I see. You've tested them before.'

'They are good hounds who run silent.'

'Such hounds have a way sooner or later of biting the hand that feeds them. They will be arrested for this burglary. The police are already after them.'

'They will take what comes to them. That is what they are paid for. I shall not appear in the matter.'

'Unless I bring you into it.'

'No, no; you would not. You are a gentleman. It is a woman's secret.'

'In the first place you must give back this manuscript.'

She broke into a ripple of laughter, and walked to the fire-place. There was a calcined* mass which she broke up with the poker. 'Shall I give this back?' she asked. So roguish and exquisite did she look as she stood before us with a challenging smile that I felt of all Holmes's criminals this was the one whom he would find it hardest to face. However, he was immune from sentiment.

'That seals your fate,' he said coldly. 'You are very prompt in your actions, madame, but you have overdone it on this occasion.'

She threw the poker down with a clatter.

'How hard you are!' she cried. 'May I tell you the whole story?'

'I fancy I could tell it to you.'

'But you must look at it with my eyes, Mr Holmes. You must realize it from the point of view of a woman who sees all her life's ambitions about to be ruined at the last moment. Is such a woman to be blamed if she protects herself?'

'The original sin was yours.'

'Yes, yes! I admit it. He was a dear boy, Douglas, but it so chanced that he could not fit into my plans. He wanted marriage—marriage, Mr Holmes—with a penniless commoner. Nothing less would serve him. Then he became pertinacious. Because I had given he seemed to think that I still must give, and to him only. It was intolerable. At last I had to make him realize it.'

'By hiring ruffians to beat him under your own window.'

'You do indeed seem to know everything. Well, it is true. Barney and the boys drove him away, and were, I admit, a little rough in doing so. But what did he do then? Could I have believed that a gentleman would do such an act? He wrote a book in which he described his own story. I, of course, was the wolf; he was the lamb. It was all there, under different names, of course; but who in all London would have failed to recognize it? What do you say to that, Mr Holmes?'

'Well, he was within his rights.'

'It was as if the air of Italy had got into his blood and brought with it the old cruel Italian spirit. He wrote to me and sent me a copy of his book that I might have the torture of anticipation. There were two copies, he said—one for me, one for his publisher.'

'How did you know the publisher's had not reached him?'

'I knew who his publisher was. It is not his only novel, you know. I found out that he had not heard from Italy. Then came Douglas's sudden death. So long as that other manuscript was in the world there was no safety for me. Of course, it must be among his effects, and these would be returned to his mother. I set the gang at work. One of them got into the house as servant. I wanted to do the thing honestly. I really and truly did. I was ready to buy the house and everything in it. I offered any price she cared to ask. I only tried the other way when everything else had failed. Now, Mr Holmes, granting that I was too hard on Douglas—and, God knows, I am sorry for it!—what else could I do with my whole future at stake?'

Sherlock Holmes shrugged his shoulders.

'Well, well,' said he, 'I suppose I shall have to compound a felony* as usual. How much does it cost to go round the world in first-class style?'

The lady stared in amazement.

'Could it be done on five thousand pounds?'

'Well, I should think so, indeed!'

'Very good. I think you will sign me a cheque for that, and I will see that it comes to Mrs Maberley. You owe her a little change of air. Meantime, lady'—he wagged a cautionary forefinger—'have a care! Have a care! You can't play with edged tools for ever without cutting those dainty hands.'

*The Blanched Soldier**

THE ideas of my friend Watson, though limited, are exceedingly pertinacious. For a long time he has worried me to write an experience of my own. Perhaps I have rather invited this persecution, since I have often had occasion to point out to him how superficial are his own accounts and to accuse him of pandering to popular taste instead of confining himself rigidly to fact and figures. 'Try it yourself, Holmes!' he has retorted, and I am compelled to admit that, having taken my pen in my hand, I do begin to realize that the matter must be presented in such a way as may interest the reader. The following case can hardly fail to do so, as it is among the strangest happenings in my collection, though it chanced that Watson had no note of it in his collection. Speaking of my old friend and biographer, I would take this opportunity to remark that if I burden myself with a companion in my various little inquiries it is not done out of sentiment or caprice, but it is that Watson has some remarkable characteristics of his own, to which in his modesty he has given small attention amid his exaggerated estimates of my own performances. A confederate who foresees your conclusions and course of action is always dangerous, but one to whom each development comes as a perpetual surprise, and to whom the future is always a closed book, is, indeed, an ideal helpmate.

I find from my notebook that it was in January 1903, just after the conclusion of the Boer War,* that I had my visit from Mr James M. Dodd, a big, fresh, sunburned, upstanding Briton. The good Watson had at that time deserted me for a wife,* the only selfish action which I can recall in our association. I was alone.

It is my habit to sit with my back to the window and to place my visitors in the opposite chair, where the light falls full upon them. Mr James M. Dodd seemed somewhat at a loss how to begin the interview. I did not attempt to help him, for his silence gave me more time for observation. I have found it wise to impress clients with a sense of power, and so I gave him some of my conclusions.

'From South Africa, sir, I perceive.'

'Yes, sir,' he answered, with some surprise.

'Imperial Yeomanry,* I fancy.'

'Exactly.'

'Middlesex Corps, no doubt.'

'That is so. Mr Holmes, you are a wizard.'

I smiled at his bewildered expression.

'When a gentleman of virile appearance enters my room with such tan upon his face as an English sun could never give, and with his handkerchief in his sleeve instead of in his pocket, it is not difficult to place him. You wear a short beard, which shows that you were not a regular. You have the cut of a riding-man. As to Middlesex, your card has already shown me that you are a stockbroker from Throgmorton Street.* What other regiment would you join?'

'You see everything.'

'I see no more than you, but I have trained myself to notice what I see. However, Mr Dodd, it was not to discuss the science of observation that you called upon me this morning. What has been happening at Tuxbury Old Park?'

'Mr Holmes—!'

'My dear sir, there is no mystery. Your letter came with that heading, and as you fixed this appointment in very pressing terms, it was clear that something sudden and important had occurred.'

'Yes, indeed. But the letter was written in the afternoon, and a good deal has happened since then. If Colonel Emsworth had not kicked me out—'

'Kicked you out!'

'Well, that was what it amounted to. He is a hard nail, is Colonel Emsworth. The greatest martinet in* the Army in his day, and it was a day of rough language, too. I couldn't have stuck the Colonel if it had not been for Godfrey's sake.'

I lit my pipe and leaned back in my chair.

'Perhaps you will explain what you are talking about.'

My client grinned mischievously.

'I had got into the way of supposing that you knew everything without being told,' said he. 'But I will give you the facts, and I hope to God that you will be able to tell me what they mean. I've been awake all night puzzling my brain, and the more I think the more incredible does it become.

'When I joined up in January, 1901—just two years ago—young Godfrey Emsworth had joined the same squadron. He was Colonel Emsworth's only son—Emsworth, the Crimean VC*—and he had the fighting blood in him, so it is no wonder he volunteered. There was not a finer lad in the regiment. We formed a friendship—the sort of friendship which can only be made when one lives the same life and shares the same joys and sorrows. He was my mate—and that means a good deal in the Army. We took the rough and the smooth together for a year of hard fighting. The he was hit with a bullet from an elephant gun in the action near Diamond Hill* outside Pretoria.* I got one letter from the hospital at Cape Town* and one from Southampton.* Since then not a word—not one word, Mr Holmes, for six months and more, and he my closest pal.

'Well, when the war was over, and we all got back, I wrote to his father and asked where Godfrey was. No answer. I waited a bit and then I wrote again. This time I had a reply, short and gruff. Godfrey had gone on a voyage round the world, and it was not likely that he would be back for a year. That was all.

'I wasn't satisfied, Mr Holmes. The whole thing seemed to me so damned unnatural. He was a good lad and he would not drop a pal like that. It was not like him. Then, again, I happened to know that he was heir to a lot of money, and also that his father and he did not always hit it off too well. The old man was sometimes a bully, and young Godfrey had too much spirit to stand it. No, I wasn't satisfied, and I determined that I would get to the root of the matter. It happened, however, that my own affairs needed a lot of straightening out, after two years' absence, and so it is only this week that I have been able to take up Godfrey's case again. But since I have taken it up I mean to drop everything in order to see it through.'

Mr James M. Dodd appeared to be the sort of person whom it would be better to have as a friend than as an enemy. His blue eyes were stern and his square jaw had set hard as he spoke.

'Well, what have you done?' I asked.

'My first move was to get down to his home, Tuxbury Old Park, near Bedford,* and to see for myself how the ground lay. I wrote to the mother, therefore—I had had quite enough of the curmudgeon of a father—and I made a clean frontal attack: Godfrey was my chum, I had a great deal of interest which I might tell her of our common experiences, I should be in the neighbourhood, would there be any objection, et cetera? In reply I had quite an amiable answer from her and an offer to put me up for the night. That was what took me down on Monday.

'Tuxbury Old Hall is inaccessible—five miles from anywhere. There was no trap* at the station, so I had to walk, carrying my suit-case, and it was nearly dark before I arrived. It is a great wandering house, standing in a considerable park. I should judge it was of all sorts of ages and styles, starting on a half-timbered Elizabethan foundation* and ending in a Victorian portico.* Inside it was all panelling and tapestry and half-effaced old pictures, a house of shadows and mystery. There was a butler,* old Ralph, who seemed about the same age as the house, and there was his wife, who might have been older. She had been Godfrey's nurse, and I had heard him speak of her as second only to his mother in his affections, so I was drawn to her in spite of her queer appearance. The mother I liked also—a gentle little white mouse of a woman. It was only the Colonel himself whom I barred.

'We had a bit of a barney* right away, and I should have walked back to the station if I had not felt that it might be playing his game for me to do so. I was shown straight into his study, and there I found him, a huge, bow-backed man with a smoky skin and a straggling grey beard,* seated behind his littered desk. A red-veined nose jutted out like a vulture's beak, and two fierce grey eyes glared at me from under tufted brows. I could understand now why Godfrey seldom spoke of his father.

' "Well, sir," said he in a rasping voice. "I should be interested to know the real reasons for this visit."

'I answered that I had explained them in my letter to his wife.

' "Yes, yes; you said that you had known Godfrey in Africa. We have, of course, only your word for that."

' "I have his letters to me in my pocket."

' "Kindly let me see them."

'He glanced at the two which I handed him, and then he tossed them back.

' "Well, what then?" he asked.

' "I was fond of your son Godfrey, sir. Many ties and memories united us. Is it not natural that I should wonder at his sudden silence and should wish to know what has become of him?"

' "I have some recollection, sir, that I had already corresponded with you and had told you what had become of him. He has gone upon a voyage round the world. His health was in a poor way after his African experiences, and both his mother and I were of opinion that complete rest and change were needed. Kindly pass that explanation on to any other friends who may be interested in the matter."

' "Certainly," I answered. "But perhaps you would have the goodness to let me have the name of the steamer and of the line by which he sailed, together with the date. I have no doubt that I should be able to get a letter through to him."

'My request seemed both to puzzle and to irritate my host. His great eyebrows came down over his eyes and he tapped his fingers impatiently on the table. He looked up at last with the expression of one who has seen his adversary make a dangerous move at chess, and has decided how to meet it.

' "Many people, Mr Dodd," said he, "would take offence at your infernal pertinacity and would think that this insistence had reached the point of damned impertinence."

' "You must put it down, sir, to my real love for your son."

' "Exactly. I have already made every allowance upon that score. I must ask you, however, to drop these inquiries. Every family has its own inner knowledge and its own motives, which cannot always be made clear to outsiders, however well-intentioned. My wife is anxious to hear something of Godfrey's past which you are in a position to tell her, but I would ask you to let the present and the future alone. Such inquiries serve no useful purpose, sir, and place us in a delicate and difficult position."

'So I came to a dead end, Mr Holmes. There was no getting past it. I could only pretend to accept the situation and register a vow inwardly that I would never rest until my friend's fate had been cleared up. It was a dull evening. We dined quietly, the three of us, in a gloomy, faded old room. The lady questioned me eagerly about her son, but the old man seemed morose and depressed. I was so bored by the whole proceeding that I made an excuse as soon as I decently could and retired to my bedroom. It was a large, bare room on the ground floor, as gloomy as the rest of the house, but after a year of sleeping upon the veldt,* Mr Holmes, one is not too particular about one's quarters. I opened the curtains and looked out into the garden, remarking that it was a fine night with a bright half-moon. Then I sat down by the roaring fire with the lamp on a table beside me, and endeavoured to distract my mind with a novel. I was interrupted, however, by Ralph, the old butler, who came in with a fresh supply of coals.

' "I thought you might run short in the night-time, sir. It is bitter weather and these rooms are cold."

'He hesitated before leaving the room, and when I looked round he was standing facing me with a wistful look upon his wrinkled face.

' "Beg your pardon, sir, but I could not help hearing what you said of young Master Godfrey at dinner. You know, sir, that my wife nursed him, and so I may say I am his foster-father. It's natural we should take an interest. And you say he carried himself well, sir?"

' "There was no braver man in the regiment. He pulled me out once from under the rifles of the Boers, or maybe I should not be here."

'The old butler rubbed his skinny hands.

' "Yes, sir, yes, that is Master Godfrey all over. He was always courageous. There's not a tree in the park, sir, that he has not climbed. Nothing would stop him. He was a fine boy—and oh, sir, he was a fine man."

'I sprang to my feet.

' "Look here!" I cried. "You say he *was*. You speak as if he were dead. What is all this mystery? What has become of Godfrey Emsworth?"

'I gripped the old man by the shoulder, but he shrank away.

' "I don't know what you mean, sir. Ask the master about Master Godfrey. He knows. It is not for me to interfere."

'He was leaving the room, but I held his arm.

' "Listen," I said. "You are going to answer one question before you leave if I have to hold you all night. Is Godfrey dead?"

'He could not face my eyes. He was like a man hypnotized. The answer was dragged from his lips. It was a terrible and unexpected one.

' "I wish to God he was!" he cried, and, tearing himself free, he dashed from the room.

'You will think, Mr Holmes, that I returned to my chair in no very happy state of mind. The old man's words seemed to me to bear only one interpretation. Clearly my poor friend had become involved in some criminal, or, at the least, disreputable, transaction which touched the family honour. That stern old man had sent his son away and hidden him from the world lest some scandal should come to light. Godfrey was a reckless fellow. He was easily influenced by those around him. No doubt he had fallen into bad hands and been misled to his ruin. It was a piteous business, if it was indeed so, but even now it was my duty to hunt him out and see if I could aid him. I was anxiously pondering the matter when I looked up, and there was Godfrey Emsworth standing before me.'*

My client had paused as one in deep emotion.

'Pray continue,' I said. 'Your problem presents some very unusual features.'

'He was outside the window, Mr Holmes, with his face pressed against the glass. I have told you that I looked out at the night. When I did so, I left the curtains partly open. His figure was framed in this gap. The window came down to the ground and I could see the whole length of it, but it was his face which held my gaze. He was deadly pale—never have I seen a man so white. I reckon ghosts may look like that; but his eyes met mine, and they were the eyes of a living man. He sprang back when he saw that I was looking at him, and he vanished into the darkness.

'There was something shocking about the man, Mr Holmes. It wasn't merely that ghastly face glimmering as white as cheese in the darkness. It was more subtle than that—something slinking, something furtive, something guilty—something very unlike the frank, manly lad that I had known. It left a feeling of horror in my mind.

'But when a man has been soldiering for a year or two with brother Boer as a playmate, he keeps his nerve and acts quickly. Godfrey had hardly vanished before I was at the window. There was an awkward catch, and I was some little time before I could throw it up.* Then I nipped through and ran down the garden path in the direction that I thought he might have taken.

'It was a long path and the light was not very good, but it seemed to me something was moving ahead of me. I ran on and called his name, but it was no use. When I got to the end of the path there were several others branching in different directions to various outhouses. I stood hesitating, and as I did so I heard distinctly the sound of a closing door. It was not behind me in the house, but ahead of me, somewhere in the darkness. That was enough, Mr Holmes, to assure me that what I had seen was not a vision. Godfrey had run away from me and he had shut a door behind him. Of that I was certain.

'There was nothing more I could do, and I spent an uneasy night turning the matter over in my mind and trying to find some theory which would cover the facts. Next day I found the Colonel rather more conciliatory, and as his wife remarked that there were some places of interest in the neighbourhood, it gave me an opening to ask whether my presence for one more night would incommode them. A somewhat grudging acquiescence from the old man gave me a clear day in which to make my observations. I was already perfectly convinced that Godfrey was in hiding somewhere near, but where and why remained to be solved.

'The house was so large and so rambling that a regiment might be hid away in it and no one the wiser. If the secret lay there, it was difficult for me to penetrate it. But the door which I had heard close was certainly not in the house. I must explore the garden and see what I could find. There was no difficulty in the way, for the old people were busy in their own fashion and left me to my own devices.

'There were several small outhouses, but at the end of the garden there was a detached building of some size—large enough for a gardener's or a gamekeeper's* residence. Could this be the place whence the sound of that shutting door had come? I approached it in a careless fashion, as though I were strolling aimlessly round the grounds. As I did so, a small, brisk, bearded man in a black coat and bowler hat—not at all the gardener type—came out of the door. To my surprise, he locked it after him and put the key in his pocket. Then he looked at me with some surprise on his face.

' "Are you a visitor here?" he asked.

'I explained that I was and that I was a friend of Godfrey's.

' "What a pity that he should be away on his travels, for he would have so liked to see me," I continued.

' "Quite so. Exactly," said he, with a rather guilty air. "No doubt you will renew your visit at some more propitious time." He passed on, but when I turned I observed that he was standing watching me, half-concealed by the laurels at the far end of the garden.

'I had a good look at the little house as I passed it, but the windows were heavily curtained, and, so far as one could see, it was empty. I might spoil my own game, and even be ordered off the premises, if I were too audacious, for I was still conscious that I was being watched. Therefore, I strolled back to the house and waited for night before I went on with my inquiry. When all was dark and quiet, I slipped out of my window and made my way as silently as possible to the mysterious lodge.

'I have said that it was heavily curtained, but now I found that the windows were shuttered as well. Some light, however, was breaking through one of them, so I concentrated my attention upon this. I was in luck, for the curtain had not been quite closed, and there was a crack in the shutter so that I could see the inside of the room. It was a cheery place enough, a bright lamp and a blazing fire. Opposite to me was seated the little man whom I had seen in the morning. He was smoking a pipe and reading a paper—'

'What paper?' I asked.

My client seemed annoyed at the interruption of his narrative.

'Can it matter?' he asked.

'It is most essential.'

'I really took no notice.'

'Possibly you observed whether it was a broad-leafed paper or of that smaller type which one associates with weeklies.'

'Now that you mention it, it was not large. It might have been *The Spectator*. However, I had little thought to spare upon such details, for a second man was seated with his back to the window, and I could swear that this second man was Godfrey. I could not see his face, but I knew the familiar slope of his shoulders. He was leaning upon his elbow in an attitude of great melancholy, his body turned towards the fire. I was hesitating as to what I should do when there was a sharp tap on my shoulder, and there was Colonel Emsworth beside me.

' "This way, sir!" said he in a low voice. He walked in silence to the house and I followed him into my own bedroom. He had picked up a time-table in the hall.

' "There is a train to London at eight-thirty," said he. "The trap will be at the door at eight."

'He was white with rage, and, indeed, I felt myself in so difficult a position that I could only stammer out a few incoherent apologies, in which I tried to excuse myself by urging my anxiety for my friend.

' "The matter will not bear discussion," said he, abruptly. "You have made a most damnable intrusion into the privacy of our family. You were here as a guest and you have become a spy. I have nothing more to say, sir, save that I have no wish ever to see you again."

'At this I lost my temper, Mr Holmes, and I spoke with some warmth.

' "I have seen your son, and I am convinced that for some reason of your own you are concealing him from the world. I have no idea what your motives are in cutting him off in this fashion, but I am sure that he is no longer a free agent. I warn you, Colonel Emsworth, that until I am assured as to the safety and well-being of my friend I shall never desist in my efforts to get to the bottom of the mystery, and I shall certainly not allow myself to be intimidated by anything which you may say or do."

'The old fellow looked diabolical, and I really thought he was about to attack me. I have said that he was a gaunt, fierce old giant, and though I am no weakling I might have been hard put to it to hold my own against him. However, after a long glare of rage he turned upon his heel and walked out of the room. For my part, I took the appointed train in the morning, with the full intention of coming straight to you and asking for your advice and assistance at the appointment for which I had already written.'

Such was the problem which my visitor laid before me. It presented, as the astute reader* will have already perceived, few difficulties in its solution, for a very limited choice of alternatives must get to the root of the matter. Still, elementary as it was, there were points of interest and novelty about it which may excuse my placing it upon record. I now proceeded, using my familiar method of logical analysis, to narrow down the possible solutions.

'The servants,' I asked; 'how many were in the house?'

'To the best of my belief there were only the old butler and his wife. They seemed to live in the simplest fashion.'

'There was no servant, then, in the detached house?'

'None, unless the little man with the beard acted as such. He seemed, however, to be quite a superior person.'

'That seems very suggestive. Had you any indication that food was conveyed from the one house to the other?'

'Now that you mention it, I did see old Ralph carrying a basket down the garden walk and going in the direction of this house. The idea of food did not occur to me at the moment.'

'Did you make any local inquiries?'

'Yes, I did. I spoke to the station-master and also to the innkeeper in the village. I simply asked if they knew anything of my old comrade, Godfrey Emsworth. Both of them assured me that he had gone for a voyage round the world. He had come home and then had almost at once started off again. The story was evidently universally accepted.'

'You said nothing of your suspicions?'

'Nothing.'

'That was very wise. The matter should certainly be inquired into. I will go back with you to Tuxbury Old Park.'

'To-day?'

It happened that at the moment I was clearing up the case which my friend Watson has described as that of the Abbey School, in which the Duke of Greyminster* was so deeply involved. I had also a commission from the Sultan of Turkey* which called for immediate action, as political consequences of the gravest kind might arise from its neglect. Therefore it was not until the beginning of the next week, as my diary records, that I was able to start forth on my mission to Bedfordshire in company with Mr James M. Dodd. As we drove to Euston* we picked up a grave and taciturn gentleman of iron-grey aspect, with whom I had made the necessary arrangements.

'This is an old friend,' said I to Dodd. 'It is possible that his presence may be entirely unnecessary, and, on the other hand, it may be essential. It is not necessary at the present stage to go further into the matter.'

The narratives of Watson have accustomed the reader, no doubt, to the fact that I do not waste words or disclose my thoughts while a case is actually under consideration. Dodd seemed surprised, but nothing more was said and the three of us continued our journey together. In the train I asked Dodd one more question which I wished our companion to hear.

'You say that you saw your friend's face quite clearly at the window, so clearly that you are sure of his identity?'

'I have no doubt about it whatever. His nose was pressed against the glass. The lamplight shone full upon him.'

'It could not have been someone resembling him?'

'No, no; it was he.'

'But you say he was changed?'

'Only in colour. His face was—how shall I describe it?—it was of a fish-belly whiteness. It was bleached.'

'Was it equally pale all over?'

'I think not. It was his brow which I saw so clearly as it was pressed against the window.'

'Did you call to him?'

'I was too startled and horrified for the moment. Then I pursued him, as I have told you, but without result.'

My case was practically complete, and there was only one small incident needed to round it off. When, after a considerable drive, we arrived at the strange old rambling house which my client had described, it was Ralph, the elderly butler, who opened the door. I had requisitioned the carriage for the day and had asked my elderly friend to remain within it unless we should summon him. Ralph, a little wrinkled old fellow, was in the conventional costume of black coat and pepper-and-salt trousers, with only one curious variant. He wore brown leather gloves, which at sight of us he instantly shuffled off, laying them down on the hall table as we passed in. I have, as my friend Watson may have remarked, an abnormally acute set of senses, and a faint but incisive scent was apparent. It seemed to centre on the hall-table. I turned, placed my hat there, knocked it off, stooped to pick it up, and contrived to bring my nose within a foot of the gloves. Yes, it was undoubtedly from them that the curious tarry odour was oozing. I passed on into the study with my case complete. Alas, that I should have to show my hand so when I tell my own story! It was by concealing such links in the chain that Watson was enabled to produce his meretricious finales.

Colonel Emsworth was not in his room, but he came quickly enough on receipt of Ralph's message. We heard his quick, heavy step in the passage. The door was flung open and he rushed in with bristling beard and twisted features, as terrible an old man as ever I have seen. He held our cards in his hand, and he tore them up and stamped on the fragments.

'Have I not told you, you infernal busybody, that you are warned off the premises? Never dare to show your damned face here again. If you enter again without my leave I shall be within my rights if I use violence. I'll shoot you, sir! By God, I will! As to you, sir,' turning upon me, 'I extend the same warning to you. I am familiar with your ignoble profession, but you must take your reputed talents to some other field. There is no opening for them here.'

'I cannot leave here,' said my client firmly, 'until I hear from Godfrey's own lips that he is under no restraint.'

Our involuntary host rang the bell.

'Ralph,' he said, 'telephone down to the county police and ask the inspector to send up two constables. Tell him there are burglars in the house.'

'One moment,' said I. 'You must be aware, Mr Dodd, that Colonel Emsworth is within his rights and that we have no legal status within his house. On the other hand, he should recognize that your action is prompted entirely by solicitude for his son. I venture to hope that, if I were allowed to have five minutes' conversation with Colonel Emsworth, I could certainly alter his view of the matter.'

'I am not so easily altered,' said the old soldier. 'Ralph, do what I have told you. What the devil are you waiting for? Ring up the police!'

'Nothing of the sort,' I said, putting my back to the door. 'Any police interference would bring about the very catastrophe which you dread.' I took out my notebook and scribbled one word upon a loose sheet. 'That', said I, as I handed it to Colonel Emsworth, 'is what has brought us here.'

He stared at the writing with a face from which every expression save amazement had vanished.

'How do you know?' he gasped, sitting down heavily in his chair.

'It is my business to know things. That is my trade.'

He sat in deep thought, his gaunt hand tugging at his straggling beard. Then he made a gesture of resignation.

'Well, if you wish to see Godfrey, you shall. It is no doing of mine, but you have forced my hand. Ralph, tell Mr Godfrey and Mr Kent that in five minutes we shall be with them.'

At the end of that time we passed down the garden path and found ourselves in front of the mystery house at the end. A small bearded man stood at the door with a look of considerable astonishment upon his face.

'This is very sudden, Colonel Emsworth,' said he. 'This will disarrange all our plans.'

'I can't help it, Mr Kent. Our hands have been forced. Can Mr Godfrey see us?'

'Yes; he is waiting inside.' He turned and led us into a large, plainly furnished front room. A man was standing with his back to the fire, and at the sight of him my client sprang forward with outstretched hand.

'Why, Godfrey, old man, this is fine!'

But the other waved him back.

'Don't touch me, Jimmie. Keep your distance. Yes, you may well stare! I don't quite look the smart Lance-Corporal Emsworth, of B Squadron, do I?'

His appearance was certainly extraordinary. One could see that he had indeed been a handsome man with clear-cut features sunburned by an African sun, but mottled in patches over this darker surface were curious whitish patches which had bleached his skin.

'That's why I don't court visitors,' said he. 'I don't mind you, Jimmie, but I could have done without your friend. I suppose there is some good reason for it, but you have me at a disadvantage.'

'I wanted to be sure that all was well with you, Godfrey. I saw you that night when you looked into my window, and I could not let the matter rest till I had cleared things up.'

'Old Ralph told me you were there, and I couldn't help taking a peep at you. I hoped you would not have seen me, and I had to run to my burrow when I heard the window go up.'

'But what in Heaven's name is the matter?'

'Well, it's not a long story to tell,' said he, lighting a cigarette. 'You remember that morning fight at Buffelsspruit, outside Pretoria, on the Eastern railway line?* You heard I was hit?'

'Yes, I heard that, but I never got particulars.'

'Three of us got separated from the others. It was very broken country, you may remember. There was Simpson—the fellow we called Baldy Simpson—and Anderson, and I. We were clearing brother Boer, but he lay low and got the three of us. The other two were killed. I got an elephant bullet through my shoulder. I stuck on to my horse, however, and he galloped several miles before I fainted and rolled off the saddle.

'When I came to myself it was nightfall, and I raised myself up, feeling very weak and ill. To my surprise there was a house close behind me, a fairly large house with a broad stoep* and many windows. It was deadly cold. You remember the kind of numb cold which used to come at evening, a deadly, sickening sort of cold, very different from a crisp healthy frost. Well, I was chilled to the bone, and my only hope seemed to lie in reaching that house. I staggered to my feet and dragged myself along, hardly conscious of what I did. I have a dim memory of slowly ascending the steps, entering a wide-opened door, passing into a large room which contained several beds, and throwing myself down with a gasp of satisfaction upon one of them. It was unmade, but that troubled me not at all. I drew the clothes over my shivering body and in a moment I was in a deep sleep.

'It was morning when I wakened, and it seemed to me that instead of coming out into a world of sanity I had emerged into some extraordinary nightmare. The African sun flooded through the big, curtainless windows, and every detail of the great, bare, whitewashed dormitory stood out hard and clear. In front of me was standing a small, dwarf-like man with a huge, bulbous head, who was jabbering excitedly in Dutch, waving two horrible hands which looked to me like brown sponges. Behind him stood a group of people who seemed to be intensely amused by the situation, but a chill came over me as I looked at them. Not one of them was a normal human being. Every one was twisted or swollen or disfigured in some strange way. The laughter of these strange monstrosities was a dreadful thing to hear.

'It seemed that none of them could speak English, but the situation wanted clearing up, for the creature with the big head was growing furiously angry and, uttering wild beast cries, he had laid his deformed hands upon me and was dragging me out of bed, regardless of the fresh flow of blood from my wound. The little monster was as strong as a bull, and I don't know what he might have done to me had not an elderly man who was clearly in authority been attracted to the room by the hubbub. He said a few stern words in Dutch and my persecutor shrank away. Then he turned upon me, gazing at me in the utmost amazement.

'"How in the world did you come here?" he asked, in amazement. "Wait a bit! I see that you are tired out and that wounded shoulder of yours wants looking after. I am a doctor, and I'll soon have you tied up. But, man alive! you are in far greater danger here than ever you were on the battlefield. You are in the Leper Hospital, and you have slept in a leper's bed."

'Need I tell you more, Jimmie? It seems that in view of the approaching battle all these poor creatures had been evacuated the day before. Then, as the British advanced, they had been brought back by this, their medical superintendent, who assured me that, though he believed he was immune to the disease, he would none the less never have dared to do what I had done. He put me in a private room, treated me kindly, and within a week or so I was removed to the general hospital at Pretoria.

'So there you have my tragedy. I hoped against hope, but it was not until I had reached home that the terrible signs which you see upon my face told me that I had not escaped. What was I to do? I was in this lonely house. We had two servants whom we could utterly trust. There was a house where I could live. Under pledge of secrecy, Mr Kent, who is a surgeon, was prepared to stay with me. It seemed simple enough on those lines. The alternative was a dreadful one—segregation for life among strangers with never a hope of release. But absolute secrecy was necessary, or even in this quiet country-side there would have been an outcry, and I should have been dragged to my horrible doom. Even you, Jimmie—even you had to be kept in the dark. Why my father has relented I cannot imagine.'

Colonel Emsworth pointed to me.

'This is the gentleman who forced my hand.' He unfolded the scrap of paper on which I had written the word 'Leprosy'. 'It seemed to me that if he knew so much as that it was safer that he should know all.'

'And so it was,' said I. 'Who knows but good may come of it? I understand that only Mr Kent has seen the patient. May I ask, sir, if you are an authority on such complaints, which are, I understand, tropical or semi-tropical in their nature?'

'I have the ordinary knowledge of the educated medical man,' he observed, with some stiffness.

'I have no doubt, sir, that you are fully competent, but I am sure that you will agree that in such a case a second opinion is valuable. You have avoided this, I understand, for fear that pressure should be put upon you to segregate the patient.'

'That is so,' said Colonel Emsworth.

'I foresaw this situation,' I explained, 'and I have brought with me a friend whose discretion may absolutely be trusted. I was able once to do him a professional service, and he is ready to advise as a friend rather than as a specialist. His name is Sir James Saunders.'

The prospect of an interview with Lord Roberts* would not have excited greater wonder and pleasure in a raw subaltern than was now reflected upon the face of Mr Kent.

'I shall indeed be proud,' he murmured.

'Then I will ask Sir James to step this way. He is at present in the carriage outside the door. Meanwhile, Colonel Emsworth, we may perhaps assemble in your study, where I could give the necessary explanations.'

And here it is that I miss my Watson. By cunning questions and ejaculations of wonder he could elevate my simple art, which is but systematized common sense,* into a prodigy. When I tell my own story I have no such aid. And yet I will give my process of thought even as I gave it to my small audience, which included Godfrey's mother, in the study of Colonel Emsworth.

'That process', said I, 'starts upon the supposition that when you have eliminated all which is impossible, then whatever remains, however improbable, must be the truth. It may well be that several explanations remain, in which case one tries test after test until one or other of them has a convincing amount of support. We will now apply this principle to the case in point. As it was first presented to me, there were three possible explanations of the seclusion or incarceration of this gentleman in an outhouse of his father's mansion. There was the explanation that he was in hiding for a crime, or that he was mad and that they wished to avoid an asylum, or that he had some disease which caused his segregation. I could think of no other adequate solutions. These, then, had to be sifted and balanced against each other.

'The criminal solution would not bear inspection. No unsolved crime had been reported from that district. I was sure of that. If it were some crime not yet discovered, then clearly it would be to the interest of the family to get rid of the delinquent and send him abroad rather than keep him concealed at home. I could see no explanation for such a line of conduct.

'Insanity was more plausible. The presence of the second person in the outhouse suggested a keeper. The fact that he locked the door when he came out strengthened the supposition and gave the idea of constraint. On the other hand, this constraint could not be severe or the young man could not have got loose and come down to have a look at his friend. You will remember, Mr Dodd, that I felt round for points, asking you, for example, about the paper which Mr Kent was reading. Had it been *The Lancet** or *The British Medical Journal** it would have helped me. It is not illegal, however, to keep a lunatic upon private premises so long as there is a qualified person in attendance and the authorities have been duly notified. Why, then, all this desperate desire for secrecy? Once again I could not get the theory to fit the facts.

'There remained the third possibility, into which, rare and unlikely as it was, everything seemed to fit. Leprosy is not uncommon in South Africa. By some extraordinary chance this youth might have contracted it. His people would be placed in a very dreadful position, since they would desire to save him from segregation. Great secrecy would be needed to prevent rumours from getting about and subsequent interference by the authorities. A devoted medical man, if sufficiently paid, would easily be found to take charge of the sufferer. There would be no reason why the latter should not be allowed freedom after dark. Bleaching of the skin is a common result of the disease. The case was a strong one—so strong that I determined to act as if it were actually proved. When on arriving here I noticed that Ralph, who carries out the meals, had gloves which are impregnated with disinfectants, my last doubts were removed. A single word showed you, sir, that your secret was discovered, and if I wrote rather than said it, it was to prove to you that my discretion was to be trusted.'

I was finishing this little analysis of the case when the door was opened and the austere figure of the great dermatologist* was ushered in. But for once his sphinx-like features had relaxed and there was a warm humanity in his eyes. He strode up to Colonel Emsworth and shook him by the hand.

'It is often my lot to bring ill-tidings, and seldom good,' said he. 'This occasion is the more welcome. It is not leprosy.'

'What?'

'A well-marked case of pseudo-leprosy or ichthyosis, a scale-like affection of the skin, unsightly, obstinate, but possibly curable, and certainly non-infective. Yes, Mr Holmes, the coincidence is a remarkable one. But is it coincidence? Are there not subtle forces at work of which we know little? Are we assured that the apprehension, from which this young man has no doubt suffered terribly since his exposure to its contagion, may not produce a physical effect which simulates that which it fears?* At any rate, I pledge my professional reputation—But the lady has fainted! I think that Mr Kent had better be with her until she recovers from this joyous shock.'*

The Lion's Mane

IT is a most singular thing that a problem which was certainly as abstruse and unusual as any which I have faced in my long professional career should have come to me after my retirement; and be brought, as it were, to my very door. It occurred after my withdrawal to my little Sussex home, when I had given myself up entirely to that soothing life of Nature for which I had so often yearned during the long years spent amid the gloom of London. At this period of my life the good Watson had passed almost beyond my ken. An occasional week-end visit was the most I ever saw of him. Thus I must act as my own chronicler. Ah! had he but been with me, how much he might have made of so wonderful a happening and of my eventual triumph against every difficulty! As it is, however, I must needs tell my tale in my own plain way, showing by my words each step upon the difficult road which lay before me as I searched for the mystery of the Lion's Mane.*

My villa is situated upon the southern slope of the Downs,* commanding a great view of the Channel. At this point the coast line is entirely of chalk cliffs, which can only be descended by a single, long, tortuous path, which is steep and slippery. At the bottom of the path lie a hundred yards of pebbles and shingle, even when the tide is at full. Here and there, however, there are curves and hollows which make splendid swimming pools filled afresh with each flow. This admirable beach extends for some miles in each direction, save only at one point where the little cove and village of Fulworth break the line.

My house is lonely. I, my old housekeeper, and my bees have the estate all to ourselves. Half a mile off, however, is Harold Stackhurst's well-known coaching establishment,* The Gables—quite a large place, which contains some score of young fellows preparing for various professions, with a staff of several masters.* Stackhurst himself was a well-known rowing Blue* in his day, and an excellent all-round scholar. He and I were always friendly from the day I came to the coast, and he was the one man who was on such terms with me that we could drop in on each other in the evenings without an invitation.

Towards the end of July 1907, there was a severe gale, the wind blowing up-Channel, heaping the seas to the base of the cliffs, and leaving a lagoon at the turn of the tide. On the morning of which I speak the wind had abated, and all Nature was newly washed and fresh. It was impossible to work upon so delightful a day, and I strolled out before breakfast to enjoy the exquisite air. I walked along the cliff path which led to the steep descent to the beach. As I walked I heard a shout behind me, and there was Harold Stackhurst waving his hand in cheery greeting.

'What a morning, Mr Holmes! I thought I should see you out.'

'Going for a swim, I see.'

'At your old tricks again,' he laughed, patting his bulging pocket. 'Yes, McPherson started early, and I expect I may find him there.'

Fitzroy McPherson was the science master, a fine upstanding young fellow whose life had been crippled by heart trouble following rheumatic fever. He was a natural athlete, however, and excelled in every game which did not throw too great a strain upon him. Summer and winter he went for his swim, and, as I am a* swimmer myself, I have often joined him.

At this moment we saw the man himself. His head showed above the edge of the cliff where the path ends. Then his whole figure appeared at the top, staggering like a drunken man. The next instant he threw up his hands, and, with a terrible cry, fell upon his face. Stackhurst and I rushed forward—it may have been fifty yards—and turned him on his back. He was obviously dying. Those glazed sunken eyes and dreadful livid cheeks could mean nothing else. One glimmer of life came into his face for an instant, and he uttered two or three words with an eager air of warning.* They were slurred and indistinct, but to my ear the last of them, which burst in a shriek from his lips,* were 'the lion's mane'. It was utterly irrelevant and unintelligible, and yet I could twist the sound into no other sense. Then* he half raised himself from the ground, threw his arms into the air* and fell forward on his side. He was dead.

My companion was paralysed by the sudden horror of it, but I, as may well be imagined, had every sense on the alert. And I had need, for it was speedily evident that we were in the presence of an extraordinary case. The man was dressed only in his Burberry overcoat,* his trousers, and an unlaced pair of canvas shoes. As he fell over, his Burberry, which had been simply thrown round his shoulders, slipped off, exposing his trunk. We stared at it in amazement. His back was covered with dark red lines as though he had been terribly flogged by a thin wire scourge. The instrument with which this punishment had been inflicted was clearly flexible, for the long, angry weals curved round his shoulders and ribs. There was blood dripping down his chin, for he had bitten through his lower lip in the paroxysm of his agony. His drawn and distorted face told how terrible that agony had been.

I was kneeling and Stackhurst standing by the body when a shadow fell across us, and we found that Ian Murdoch was by our side. Murdoch was the mathematical coach at the establishment, a tall, dark, thin man, so taciturn and aloof that none can be said to have been his friend. He seemed to live in some high abstract region of surds and conic sections* with little to connect him with ordinary life. He was looked upon as an oddity by the students, and would have been their butt, but there was some strange outlandish blood in the man which showed itself not only in his coal-black eyes and swarthy face, but also in occasional outbreaks of temper, which could only be described as ferocious. On one occasion, being plagued by a little dog belonging to McPherson, he had caught the creature up and hurled it through the plate-glass window—an action for which Stackhurst would certainly have given him his dismissal had he not been a very valuable teacher. Such was the strange, complex man who now appeared beside us. He seemed to be honestly shocked at the sight before him, though the incident of the dog may show that there was no great sympathy between the dead man and himself.

'Poor fellow! Poor fellow! What can I do? How can I help?'

'Were you with him? Can you tell us what has happened?'

'No, no, I was late this morning. I was not on the beach at all. I have come straight from The Gables. What can I do?'*

'You can hurry to the police station at Fulworth. Report the matter at once.'

Without a word he made* off at top speed, and I proceeded to take the matter in hand, while Stackhurst, dazed at this tragedy, remained by the body.* My first task naturally was to note who was on the beach. From the top of the path I could see the whole sweep of it, and it was absolutely deserted save that two or three dark figures could be seen far away moving towards the village of Fulworth. Having satisfied myself upon this point, I walked slowly down the path. There was clay or soft marl mixed with the chalk, and every here and there I saw the same footstep, both ascending and descending. No one else had* gone down to the beach by this track* that morning. At one place I observed the print of an open hand with the fingers towards the incline. This could only mean that poor McPherson had fallen as he ascended. There were rounded depressions, too, which suggested that he had come down upon his knees more than once. At the bottom of the path was the considerable lagoon left by the retreating tide. At the side of it McPherson had undressed, for there lay* his towel on a rock. It was folded* and dry, so that it would seem* that after all he had never entered the water.* Once or twice as I hunted round* amid the hard shingle I came on little patches of sand where the print* of his canvas shoe, and also of his naked foot, could be seen. The latter fact proved that he had made all ready to bathe, though the towel indicated* that he had not actually done so.*

And here was the problem clearly defined—as strange a one as had ever confronted me.* The man had not been on the beach more than a quarter of an hour at the most. Stackhurst had followed him from The Gables, so there could be no doubt about that. He had gone to bathe and had stripped, as the naked* footsteps showed. Then he had suddenly* huddled on his clothes again—they were all dishevelled and unfastened—and he had returned without bathing, or at any rate without drying himself.* And the reason for his change of purpose had been that he had been scourged in some savage, inhuman fashion, tortured until he bit his lip through in his agony, and was* left with only strength enough to crawl away and to die. Who had done this barbarous deed? There were, it is true, small grottoes and caves in the base of the cliffs, but the low sun shone directly into them, and there was no place for concealment.* Then, again,* there were those distant figures on the beach. They seemed too far away to have been connected with the crime, and the broad lagoon in which McPherson had intended to bathe lay between him and them, lapping up to the rocks. On the sea two or three fishing boats were* at no great distance. Their occupants might be examined at our leisure. There were several roads* for inquiry, but none which led to any very obvious* goal.

When I at last returned to the body I found that a little group of wandering folk had gathered round it. Stackhurst was, of course,* still there, and Ian Murdoch had just arrived with Anderson, the village constable, a big, ginger-moustached man of the slow solid Sussex breed—a breed which covers much good sense under a heavy, silent exterior. He listened to everything, took note of all we said, and finally drew* me aside.

'I'd be glad of your advice, Mr Holmes. This is a big thing for me to handle, and I'll hear of it from Lewes* if I go wrong.'

I advised him to send for his immediate superior, and for a doctor; also* to allow nothing to be removed, and as few fresh* footmarks as possible to be made,* until they came. In the mean time I searched the dead man's pockets. There were his handkerchief, a large knife,* and a small folding card-case. From this projected a slip of paper, which I unfolded and handed to the constable. There was written on it in a scrawling feminine hand: 'I will be there you may be sure. Maudie.' It read like a love affair, an assignation, though when and where were a blank. The constable replaced it in the card-case and returned it with the other things to the pockets of the Burberry. Then, as nothing more suggested itself, I walked back to my house for breakfast, having first arranged that the base of the cliffs should be thoroughly searched.*

Stackhurst was round in an hour or two to tell me that the body had been removed to* The Gables, where the inquest would be held. He brought with him some serious and definite news. As I expected,* nothing had been found in the small caves below the cliff, but he* had examined the papers in* McPherson's desk, and there were several which showed an intimate correspondence with a certain Miss Maud Bellamy, of Fulworth.* We had then established the identity of the writer of the note.

'The police have the letters,'* he explained. 'I could not bring them. But there is no doubt that it was a serious love affair. I see no reason, however, to connect it with that horrible happening save, indeed,* that the lady had made an appointment with him.'

'But hardly at a bathing-pool which all of you* were in the habit of using,' I remarked.*

'It is mere chance', said he, 'that several of the* students were not with McPherson.'

'*Was* it mere chance?'

Stackhurst knit his brows in thought.

'Ian Murdoch held them back,' said he, 'he would insist upon some algebraic* demonstration before breakfast. Poor chap, he is dreadfully cut up about it all.'

'And yet I gather that they were not friends.'

'At one time they were not. But for a year or more Murdoch has been as near to McPherson as he ever could be to anyone. He is not of a very sympathetic disposition by nature.'

'So I understand. I seem to* remember your telling me once about a* quarrel over the ill-usage of a dog.'

'That blew over all right.'

'But left some vindictive feeling, perhaps.'

'No, no; I am sure they were real friends.'

'Well, then, we must explore the matter of the girl. Do you know her?'

'Everybody knows her. She is the beauty of the neighbourhood—a real beauty, Holmes, who would draw* attention everywhere. I knew that McPherson was attracted by her, but I had no notion that it had gone so far as these letters would seem to indicate.'

'But who is she?'

'She is the daughter of old Tom Bellamy, who owns* all the boats and bathing-cots* at Fulworth. He was a fisherman to start with, but is now a man of some substance. He and his son William run the business.'

'Shall we walk into Fulworth* and see them?'

'On what pretext?'

'Oh, we can easily find a pretext. After all, this poor man did not ill-use* himself in this outrageous way. Some human hand was on the handle of that scourge, if indeed it was a scourge* which inflicted the injuries. His circle of acquaintances in this lonely place was surely limited. Let us follow it up in every direction and we can hardly fail to come upon the motive, which in turn should lead us to the criminal.'

It would have been a pleasant walk across the thyme-scented downs had our minds not been poisoned by the tragedy we had witnessed. The village of Fulworth lies in a hollow curving in a semicircle round the bay.* Behind the old-fashioned hamlet several modern houses have been built upon the rising* ground. It was to one of these that Stackhurst guided me.

'That's The Haven, as Bellamy called it. The one with the corner tower and slate roof. Not bad for a man who started with nothing but—By Jove, look at that!'

The garden gate of The Haven had opened and a man had emerged. There was no mistaking that tall, angular, straggling figure. It was Ian Murdoch, the mathematician. A moment later we confronted him upon the road.

'Hullo!' said Stackhurst. The man nodded, gave us a sideways* glance from his curious dark eyes, and would have passed* us, but his principal pulled him up.

'What were you doing there?' he asked.

Murdoch's face flushed with anger. 'I am your subordinate, sir, under your roof. I am not aware that I owe you any account of my private actions.'

Stackhurst's nerves were near the surface after all he had endured. Otherwise, perhaps, he would have waited. Now he lost his temper completely.

'In the circumstances your answer is pure impertinence, Mr Murdoch.'

'Your own question might perhaps come under the same heading.'

'This is not the first time that I have had to overlook your insubordinate ways. It will certainly be the last. You will kindly make fresh arrangements for your future as speedily as you can.'

'I had intended to do so. I have lost to-day the only person who made The Gables habitable.'*

He strode off upon his way, while Stackhurst, with angry eyes,* stood glaring after him. 'Is he not an impossible, intolerable man?' he cried.

The one thing that impressed itself forcibly* upon my mind was that Mr Ian Murdoch was taking the first chance to open a path of escape from the scene of the crime. Suspicion, vague and nebulous, was now beginning to take outline in my mind. Perhaps the visit to the Bellamys might throw some further light upon the matter. Stackhurst pulled himself together and we went forward to the house.

Mr Bellamy proved to be a middle-aged man with a flaming red beard. He seemed to be in a very angry mood, and his face was soon as florid as his hair.

'No, sir, I do not desire any particulars. My son here'—indicating a powerful young man, with a heavy, sullen face, in the corner of the sitting-room—'is of one mind with me that Mr McPherson's attentions to Maud were insulting. Yes, sir, the word "marriage" was never mentioned, and yet there were letters and meetings, and a great deal more of which neither of us could approve. She has no mother, and we are her only guardians. We are determined—'

But the words were taken from his mouth by the appearance of the lady herself. There was no gainsaying that she would have graced any assembly in the world. Who could have imagined that so rare a flower would grow from such a root and in such an atmosphere? Women have seldom been an attraction to me, for my brain has always governed my heart, but I could not look upon her perfect clear-cut face, with all the soft freshness of the Downlands in her delicate colouring, without realizing that no young man would cross her path unscathed. Such was the girl who had pushed open the door and stood now, wide-eyed and intense,* in front of Harold Stackhurst.

'I know already* that Fitzroy* is dead,' she said. 'Do not be afraid to tell me the particulars.'

'This other gentleman of yours let us know the news,' explained the father.

'There is no reason why my sister should be brought into the matter,' growled the younger man.

The sister turned a sharp, fierce look upon him. 'This is my business, William. Kindly leave me to manage it in my own way. By all accounts there has been a crime committed. If I can help to show who did it, it is the least I can do for him who is gone.'

She listened to a short account from my companion, with a composed concentration* which showed me that she possessed strong character as well as great beauty. Maud Bellamy will always remain in my memory as a most complete and remarkable woman. It seems* that she already knew me by sight, for she turned to me at the end.

'Bring them to justice, Mr Holmes. You have my sympathy and my help, whoever they may be.' It seemed to me that she glanced defiantly at her father and brother as she spoke.

'Thank you,' said I. 'I value a woman's instinct in such matters. You use the word "they". You think that more than one person was concerned?'

'I knew Mr McPherson well enough to be aware that he was a brave and a strong man. No single person could ever have inflicted such an outrage upon him.'

'Might I have one word with you alone?'

'I tell you, Maud, not to mix yourself up in the matter,' cried her father angrily.

She looked at me helplessly. 'What can I do?'

'The whole world will know the facts presently, so there can be no harm if I discuss them here,'* said I. 'I should have preferred privacy, but if your father will not allow it, he must share the deliberations.' Then I spoke of the note which had been found in the dead man's pocket. 'It is sure to be produced at the inquest. May I ask you to throw any light upon it that you can?'

'I see no reason for mystery,' she answered. 'We were engaged to be married, and we only kept it secret because Fitzroy's* uncle, who is very old and said to be dying, might have disinherited him if he had married against his wish. There was no other reason.'

'You could have told us,' growled Mr Bellamy.

'So I would, father, if you had ever shown sympathy.'

'I object to my girl picking up with men outside her own station.'

'It was your prejudice against him which prevented us from* telling you. As to this appointment'—she fumbled in her dress and produced a crumpled note—'it was in answer to this.'

Dearest, [ran the message]

The old place on the beach just after sunset on Tuesday. It is the only time I can get away.—F. M.

'Tuesday was to-day, and I had meant to meet him tonight.'

I turned over the paper. 'This never came by post. How did you get it?'

'I would rather not answer that question. It has really nothing to do with the matter which you are investigating. But anything* which bears upon that I will most freely answer.'

She was as good as her word,* but there was nothing which was helpful in our investigation. She had no reason to think that her *fiancé* had any hidden enemy, but she admitted* that she had had several warm admirers.

'May I ask if Mr Ian Murdoch was one of them?'

She blushed and seemed confused.

'There was a time when I thought he was. But that was all changed when he understood the relations between Fitzroy* and myself.'

Again the shadow round this strange man seemed to me to be taking more definite shape. His record must be examined. His rooms must be privately searched. Stackhurst was a willing collaborator, for in his mind also suspicions were forming. We returned from our visit to The Haven with the hope that one free end of this tangled skein was already* in our hands.

A week* passed. The inquest had thrown no light upon the matter and had been adjourned for further evidence. Stackhurst had made discreet inquiry about his subordinate, and there had been a superficial search of his room,* but without result. Personally, I had gone over the whole ground again, both physically and mentally, but with no new conclusions.* In all my chronicles the reader will find no case which brought me so completely to the limit of my powers. Even my imagination could conceive no solution to the mystery. And then there came the incident of the dog.*

It was my old housekeeper who heard of it first by that strange wireless by which such people collect the news of the countryside.

'Sad story this, sir, about Mr McPherson's dog,' said she one evening.

I do not encourage such conversations, but the words arrested my attention.

'What of Mr McPherson's dog?'

'Dead, sir. Died of grief for its master.'

'Who told you this?'

'Why, sir, everyone is talking of it. It took on terrible, and has eaten nothing for a week. Then to-day two of the young gentlemen from The Gables found it dead—down on the beach, sir, at the very place where its master met his end.'

'At the very place.' The words stood out clear in my memory. Some dim perception that the matter was vital rose in my mind. That the dog should die was after the beautiful, faithful* nature of dogs. But 'in the very place'! Why should this lonely beach be fatal to it? Was it possible that it also, had been sacrificed to some revengeful feud? Was it possible—Yes, the perception was dim, but already something was building up in my mind. In a few minutes I was on my way to The Gables, where I found Stackhurst in his study. At my request he sent for Sudbury and Blount, the two students who had found the dog.

'Yes, it lay on the very edge of the pool,' said one of them. 'It must have followed the trail of its dead master.'

I saw the faithful little creature, an Airedale terrier,* laid out upon the mat in the hall. The body was stiff and rigid, the eyes projecting, and the limbs contorted. There was agony in every line of it.

From The Gables I walked down to the bathing pool. The sun had sunk and the shadow of the great cliff lay black across the water, which glimmered dully like a sheet of lead. The place was deserted and there was no sign of life save for two sea-birds circling and screaming overhead. In the fading light I could dimly make out the little dog's spoor upon the sand round the very rock on which his master's towel had been laid. For a long time I stood in deep meditation while the shadows grew darker around me. My mind was filled with racing thoughts. You have known what it was to be in a nightmare in which you feel that there is some all-important thing for which you search and which you know is there, though it remains for ever just beyond your reach. That was how I felt that evening as I stood alone by that place of death. Then at last I turned and walked slowly homewards.

I had just reached the top of the path when it came to me. Like a flash, I remembered the thing for which I had so eagerly and vainly grasped. You will know, or Watson has written in vain, that I hold a vast store of out-of-the-way knowledge, without scientific system, but very available for the needs of my work. My mind is like a crowded boxroom with packets of all sorts stowed away therein—so many that I may well have but a vague perception of what was there. I had known that there was something which might bear upon this matter. It was still vague, but at least I knew how I could make it clear. It was monstrous, incredible, and yet it was always a possibility. I would test it to the full.

There is a great garret in my little house which is stuffed with books. It was into this that I plunged and rummaged for an hour. At the end of that time I emerged with a little chocolate and silver volume. Eagerly I turned up the chapter of which I had a dim remembrance. Yes, it was indeed a far-fetched and unlikely proposition, and yet I could not* be at rest until I had made sure if it might, indeed, be so. It was late when I retired, with my mind eagerly awaiting the work of the morrow.

But that work met with an annoying interruption. I had hardly swallowed my early cup of tea and was starting for the beach when I had a call from Inspector Bardle of the Sussex Constabulary—a steady, solid, bovine man with thoughtful eyes, which looked at me now with a very troubled expression.

'I know your immense experience sir,' said he. 'This is quite unofficial, of course, and need go no farther. But I am fairly up against it in this McPherson case. The question is, shall I make an arrest, or shall I not?'

'Meaning Mr Ian Murdoch?'*

'Yes, sir. There is really no one else when you come to think of it. That's the advantage of this solitude. We narrow it down to a very small compass. If he did not do it, then who did?'

'What have you against him?'

He had gleaned along the same furrows as I had. There was Murdoch's character and the mystery which seemed to hang round the man. His furious bursts of temper, as shown in the incident of the dog. The fact that he had quarrelled with McPherson in the past, and that there was some reason to think that he might have resented his attentions to Miss Bellamy. He had all my points, but no fresh ones, save that Murdoch seemed to be making every preparation for departure.

'What would my position be if I let him slip away with all this evidence against him?' The burly, phlegmatic man was sorely troubled in his mind.*

'Consider', I said, 'all the essential gaps in your case. On the morning of the crime he can surely prove an *alibi*. He had been with his scholars till the last moment, and within a few minutes of McPherson's appearance he came upon us from behind. Then bear in mind* the absolute impossibility that he could single-handed have inflicted this outrage upon a man quite as strong as himself. Finally* there is the question of the instrument with which these injuries were inflicted.'*

'What could it be* but a scourge or flexible whip of some sort?'

'Have you examined the marks?' I asked.

'I have seen them. So has the doctor.'*

'But I have examined them very carefully with a lens. They have peculiarities.'

'What are they, Mr Holmes?'

I stepped to my bureau and brought out an enlarged photograph. 'This is my method in such cases,' I explained.

'You certainly do things thoroughly, Mr Holmes.'

'I should hardly be what I am if I did not. Now let us consider this weal which extends round the right shoulder. Do you observe nothing remarkable?'

'I can't say I do.'

'Surely it is evident that it is unequal in its intensity. There is a dot of extravasated* blood here, and another there. There are similar indications in this other weal down here. What can that mean?'

'I have no idea. Have you?'*

'Perhaps I have. Perhaps I haven't.* I may be able to say more soon. Anything* which will define what made that mark will bring us a long way towards the criminal.'*

'It is, of course, an absurd idea,' said the policeman, 'but if a red-hot net of wire had been laid across the back, then these better-marked points would represent where the meshes crossed each other.'

'A most ingenious comparison.* Or shall we say a very stiff cat-o'-nine-tails* with small hard knots upon it?'

'By jove, Mr Holmes, I think you have hit it.'

'Or* there may be some very different cause, Mr Bardle.* But you case is far too weak for an arrest.* Besides, we have those last words—"Lion's Mane".'

'I have wondered whether Ian—'

'Yes, I have considered that. If the second word had any resemblance to Murdoch—but it did not. He gave it almost in a shriek. I am sure that it was "Mane".'*

'Have you no alternative, Mr Holmes?'

'Perhaps I have. But I do not care to discuss it until there is something solid to discuss.'

'And when will that be?'

'In an hour—possibly less.'

The Inspector rubbed his chin and looked at me with dubious eyes.

'I wish I could see what was in your mind, Mr Holmes. Perhaps it's those fishing-boats.'

'No, no; they were too far out.'

'Well, then, is it Bellamy and that big son of his? They were not too sweet upon Mr McPherson. Could they have done him a mischief?'

'No, no; you won't draw me until I am ready,' said I with a smile. 'Now, Inspector, we each have our own work to do. Perhaps if you were to meet me here at midday—?'*

So far we had got when there came the tremendous interruption which was the beginning of the end.

My outer door was flung open, there were blundering footsteps in the passage, and Ian Murdoch staggered into the room, pallid, dishevelled, his clothes in wild disorder, clawing with his bony hands at the furniture to hold himself erect. 'Brandy! Brandy!' he gasped, and fell groaning upon the sofa.

He was not alone. Behind him came Stackhurst,* hatless and panting, almost as *distrait* as his companion.

'Yes, yes, brandy!' he cried. 'The man is at his last gasp. It was all I could do to bring him here. He fainted twice upon the way.'

Half a tumbler of the raw spirit brought about a wondrous change. He pushed himself up on one arm and swung his coat off his shoulders. 'For God's sake!* oil, opium, morphia!' he cried. 'Anything to ease this infernal agony!'

The Inspector and I cried out at the sight.* There, criss-crossed upon the man's naked shoulder, was the same strange reticulated* pattern of red, inflamed lines which had been the death-mark* of Fitzroy McPherson.

The pain was evidently terrible and was more than local, for the sufferer's breathing would stop for a time, his face would turn black, and then with loud gasps he would clap his hand to his heart, while his brow dropped beads of sweat. At any moment he might die. More and more brandy was poured down his throat, each fresh dose bringing him back to life. Pads of cotton wool soaked in salad oil seemed to take the agony from the strange wounds. At last his head fell heavily upon the cushion. Exhausted Nature had taken refuge in its last storehouse of vitality. It was half a sleep and half a faint, but at least it was ease from pain.*

To question him had been impossible, but the moment we were assured of his condition Stackhurst turned upon me.*

'My God!' he cried, 'what is it, Holmes? What is it?'*

'Where did you find him?'

'Down on the beach. Exactly where poor McPherson met his end. If this man's heart had been weak as McPherson's was, he would not be here now. More than once I thought he was gone as I brought him up. It was too far to The Gables, so I made for you.'

'Did you see him on the beach?'

'I was walking on the cliff when I heard his cry. He was at the edge of the water, reeling about like a drunken man. I ran down, threw some clothes over* him and brought him up. For heaven's sake, Holmes, use all the powers you have and spare no pains to lift the curse from this place, for life is becoming unendurable. Can you, with all your world-wide reputation, do nothing for us?'

'I think I can, Stackhurst. Come with me now! And you, Inspector, come along! We will see if we cannot deliver this murderer into your hands.'

Leaving the unconscious man in the charge of my housekeeper, we all three went down to the deadly lagoon. On the shingle there was piled a little heap of towels and clothes, left by the stricken man. Slowly I walked round the edge of the water, my comrades in Indian file* behind me. Most of the pool was quite shallow, but under the cliff where the beach was hollowed out it was four or five feet deep. It was to this part that a swimmer would naturally go, for it formed a beautiful pellucid green pool as clear as crystal. A line of rocks lay above it at the base of the cliff, and along this I led the way, peering eagerly into the depths beneath me. I had* reached the deepest and stillest pool when my eyes caught that for which they were searching, and I burst into a shout of triumph.*

'Cyanea!'* I cried.* 'Cyanea! Behold the Lion's Mane!'

The strange object at which I pointed* did indeed look like a tangled mass torn from the mane of a lion. It lay upon a rocky* shelf some three feet under the water, a curious waving, vibrating, hairy creature with streaks of silver among its yellow tresses. It pulsated with a slow, heavy dilation and contraction.*

'It has done mischief enough. Its day* is over!' I cried. 'Help me, Stackhurst! Let us end the murderer for ever.'*

There was a big boulder just above the ledge, and we pushed it until it fell with a tremendous splash into the water. When the ripples had cleared we saw that it had settled upon the ledge below. One flapping* edge of yellow membrane showed that our victim was beneath it. A thick oily* scum oozed out from below the stone and stained the water round, rising slowly to the surface.

'Well, this gets me!' cried the Inspector. 'What was it, Mr Holmes? I'm born and bred in these parts, but I never saw such a thing. It don't belong to Sussex.'

'Just as well for Sussex,' I remarked. 'It may have been the South-west gale that brought it up.* Come back to my house, both of you, and I will give you the terrible experience of one who had good reason to remember his own meeting with the same peril of the seas.'

When we reached my study,* we found that Murdoch was so far recovered that he could sit up. He was dazed in mind, and every now and then was shaken by a paroxysm of pain. In broken words he explained that he had no notion what had occurred to him, save that terrific pangs* had suddenly shot through him, and that it had taken all his fortitude to reach the bank.

'Here is a book,' I said, taking up the little volume, 'which first brought light into what might have been for ever dark.* It is *Out of Doors*, by the famous observer J. G. Wood. Wood* himself very nearly perished from contact with this vile creature, so he wrote with a very full knowledge. *Cyanea capillata* is the miscreant's full name, and he can be as dangerous to life as, and far more painful than, the bite of the cobra.* Let me briefly give this extract.

' "If the bather should see a loose roundish mass of tawny membranes and fibres, something like very large handfuls of lion's mane and silver paper, let him beware, for this is the fearful stinger, *Cyanea capillata*." Could our sinister acquaintance be more clearly described?*

'He goes on to tell his own encounter* with one when swimming off the coast of Kent. He found that the creature radiated almost invisible filaments to the distance of fifty feet, and that anyone within that circumference from the deadly centre was in danger of death. Even at a distance the effect upon Wood was almost fatal. "The multitudinous threads caused light scarlet lines upon the skin which on closer examination resolved into minute dots or pustules, each dot charged as it were with a red-hot needle making its way through the nerves."

'The local pain was, as he explains, the least part of the exquisite torment. "Pangs shot through the chest, causing me to fall if as struck by a bullet. The pulsation would cease, and then the heart would give six or seven leaps as if it would force its way through the chest."

'It nearly killed him, although he had only been exposed to it in the disturbed* ocean and not in the narrow calm waters of a bathing pool. He says that he could hardly recognize himself afterwards, so white, wrinkled and shrivelled was his face. He gulped down brandy, a whole bottleful, and it seems to have saved his life. There is the book, Inspector.* I leave it with you, and you cannot doubt that it contains a full explanation of the tragedy of poor McPherson.'

'And incidentally exonerates me,' remarked Ian Murdoch with a wry smile. 'I do not blame you, Inspector, nor you, Mr Holmes, for your suspicions were natural. I feel that on the very eve of my arrest I have only* cleared myself by sharing the fate of my poor friend.'

'No, Mr Murdoch. I was already upon the track, and had I been out as early as I intended I might well have saved you from this terrific experience.'

'But how did you know, Mr Holmes?'

'I am an omnivorous reader with a strangely retentive memory for trifles. That phrase "Lion's Mane" haunted my mind. I knew that I had seen it somewhere in an unexpected context. You have seen that it does describe the creature. I have no doubt that it was floating on the water when McPherson saw it, and that this phrase was the only one by which he could convey to us a warning as to the creature which had been his death.'

'Then I, at least, am cleared,' said Murdoch, rising slowly to his feet. 'There are one or two words of explanation which I should give, for I know the direction in which your inquiries have run. It is true that I loved this lady, but from the day when she chose my friend McPherson my one desire was to help her to happiness. I was well content to stand aside and act as their go-between. Often I carried their messages, and it was because I was in their confidence and because she was so dear to me that I hastened to tell her of my friend's death, lest someone should forestall me in a more sudden and heartless manner. She would not tell you, sir, of our relations lest you would disapprove and I might suffer. But with your leave I must try to get back to The Gables, for my bed will be very welcome.'

Stackhurst held out his hand. 'Our nerves have all been at concert pitch,'* said he. 'Forgive what is past, Murdoch. We shall understand each other better in the future.' They passed out together with their arms linked in friendly fashion. The Inspector remained, staring at me in silence with his ox-like eyes.

'Well, you've done it!' he cried at last. 'I had read of you, but I never believed it. It's wonderful!'

I was forced to shake my head. To accept such praise was to lower one's own standards.

'I was slow at the outset—culpably slow. Had the body been found in the water I could hardly have missed it. It was the towel which misled me. The poor fellow had never thought to dry himself, and so I in turn was led to believe that he had never been in the water. Why, then, should the attack of any water creature suggest itself to me? That was where I went astray. Well, well, Inspector, I often ventured* to chaff you gentlemen of the police force, but *Cyanea capillata* very nearly avenged Scotland Yard.'*

The Retired Colourman

SHERLOCK HOLMES was in a melancholy and philosophic mood that morning. His alert practical nature was subject to such reactions.

'Did you see him?' he asked.

'You mean the old fellow who has just gone out?'

'Precisely.'

'Yes, I met him at the door.'

'What did you think of him?'

'A pathetic, futile, broken creature.'

'Exactly, Watson. Pathetic and futile. But is not all life pathetic and futile? Is not his story a microcosm of the whole? We reach. We grasp. And what is left in our hands at the end? A shadow. Or worse than a shadow—misery.'

'Is he one of your clients?'

'Well, I suppose I may call him so. He has been sent on by the Yard. Just as medical men occasionally send their incurables to a quack. They argue that they can do nothing more, and that whatever happens the patient can be no worse than he is.'

'What is the matter?'

Holmes took a rather soiled card from the table. 'Josiah Amberley. He says he was junior partner of Brickfall and Amberley, who are manufacturers of artistic materials. You will see their names upon paint-boxes. He made his little pile, retired from business at the age of sixty-one, bought a house at Lewisham* and settled down to rest after a life of ceaseless grind. One would think his future was tolerably assured.'

'Yes, indeed.'

Holmes glanced over some notes which he had scribbled upon the back of an envelope.

'Retired in 1896, Watson. Early in 1897 he married a woman twenty years younger than himself—a good-looking woman, too, if the photograph does not flatter. A competence, a wife, leisure—it seemed a straight road which lay before him. And yet within two years* he is, as you have seen, as broken and miserable a creature as crawls beneath the sun.'

'But what has happened?'

'The old story, Watson. A treacherous friend and a fickle wife. It would appear that Amberley has one hobby in life, and it is chess. Not far from him at Lewisham there lives a young doctor who is also a chess-player. I have noted his name as Dr Ray Ernest. Ernest was frequently in the house, and an intimacy between him and Mrs Amberley was a natural sequence, for you must admit that our unfortunate client has few outward graces, whatever his inner virtues may be. The couple went off together last week—destination untraced. What is more, the faithless spouse carried off the old man's deed-box as her personal luggage with a good part of his life's savings within. Can we find the lady? Can we save the money? A commonplace problem so far as it has developed, and yet a vital one for Josiah Amberley.'

'What will you do about it?'

'Well, the immediate question, my dear Watson, happens to be, What will *you* do?—if you will be good enough to understudy me. You know that I am preoccupied with this case of the two Coptic Patriarchs,* which should come to a head to-day. I really have not time to go out to Lewisham, and yet evidence taken on the spot has a special value. The old fellow was quite insistent that I should go, but I explained my difficulty. He is prepared to meet a representative.'

'By all means,' I answered. 'I confess I don't see that I can be of much service, but I am willing to do my best.' And so it was that on a summer afternoon I set forth to Lewisham, little dreaming that within a week the affair in which I was engaging would be the eager debate of all England.

It was late that evening* before I returned to Baker Street and gave an account of my mission. Holmes lay with his gaunt figure stretched in his deep chair, his pipe curling forth slow wreaths of acrid tobacco, while his eyelids drooped over his eyes so lazily that he might almost have been asleep were it not that at any halt or questionable passage of my narrative they half lifted, and two grey eyes, as bright and keen as rapiers, transfixed me with their searching glance.

'The Haven is the name of Mr Josiah Amberley's house,' I explained. 'I think it would interest you, Holmes. It is like some penurious patrician who has sunk into the company of his inferiors. You know that particular quarter, the monotonous brick streets, the weary suburban highways. Right in the middle of them, a little island of ancient culture and comfort, lies this old home, surrounded by a high sun-baked wall, mottled with lichens and topped with moss, the sort of wall—'

'Cut out the poetry, Watson,' said Holmes, severely. 'I note that it was a high brick wall.'

'Exactly. I should not have known which was The Haven had I not asked a lounger who was smoking in the street. I have reason for mentioning him. He was a tall, dark, heavily-moustached, rather military-looking man. He nodded in answer to my inquiry and gave me a curiously questioning glance, which came back to my memory a little later.

'I had hardly entered the gateway before I saw Mr Amberley coming down the drive. I only had a glimpse of him this morning, and he certainly gave me the impression of a strange creature, but when I saw him in full light his appearance was even more abnormal.'

'I have, of course, studied it, and yet I should be interested to have your impression,' said Holmes.

'He seemed to me like a man who was literally bowed down by care. His back was curved as though he carried a heavy burden. Yet he was not the weakling that I had at first imagined, for his shoulders and chest have the framework of a giant, though his figure tapers away into a pair of spindled legs.'

'Left shoe wrinkled, right one smooth.'

'I did not observe that.'

'No, you wouldn't. I spotted his artificial limb. But proceed.'

'I was struck by the snaky locks of grizzled hair which curled from under his old straw hat, and his face with its fierce, eager expression and the deeply-lined features.'

'Very good, Watson. What did he say?'

'He began pouring out the story of his grievances. We walked down the drive together, and of course I took a good look round. I have never seen a worse-kept place. The garden was all running to seed, giving me an impression of wild neglect in which the plants had been allowed to find the way of nature rather than of art. How any decent woman could have tolerated such a state of things, I don't know. The house, too, was slatternly to the last degree, but the poor man seemed himself to be aware of it and to be trying to remedy it, for a great pot of green paint stood in the centre of the hall and he was carrying a thick brush in his left hand. He had been working on the woodwork.

'He took me into his dingy sanctum, and we had a long chat. Of course, he was disappointed that you had not come yourself. "I hardly expected," he said, "that so humble an individual as myself, especially after my heavy financial loss, could obtain the complete attention of so famous a man as Mr Sherlock Holmes."

'I assured him that the financial question did not arise. "No, of course, it is art for art's sake with him," said he; "but even on the artistic side of crime he might have found something here to study. And human nature, Dr Watson—the black ingratitude of it all! When did I ever refuse one of her requests? Was ever a woman so pampered? And that young man—he might have been my own son. He had the run of my house. And yet see how they have treated me! Oh, Dr Watson, it is a dreadful, dreadful world!"

'That was the burden of his song for an hour or more. He had, it seems, no suspicion of an intrigue. They lived alone save for a woman who comes in by the day and leaves every evening at six. On that particular evening old Amberley, wishing to give his wife a treat, had taken two upper circle seats at the Haymarket Theatre.* At the last moment she had complained of a headache and had refused to go. He had gone alone. There seemed to be no doubt about the fact, for he produced the unused ticket which he had taken for his wife.'

'That is remarkable—most remarkable,' said Holmes, whose interest in the case seemed to be rising. 'Pray continue, Watson. I find your narrative most arresting. Did you personally examine this ticket? You did not, perchance, take the number?'

'It so happens that I did,' I answered with some pride. 'It chanced to be my old school number,* thirty-one, and so it stuck in my head.'

'Excellent, Watson! His seat, then, was either thirty or thirty-two.'

'Quite so,' I answered, with some mystification. 'And on B row.'

'That is most satisfactory. What else did he tell you?'

'He showed me his strong-room, as he called it. It really is a strong-room—like a bank—with iron door and shutter—burglar-proof, as he claimed. However, the woman seems to have had a duplicate key, and between them they had carried off some seven thousand pounds' worth of cash and securities.'

'Securities! How could they dispose of those?'

'He said that he had given the police a list and that he hoped they would be unsaleable. He had got back from the theatre about midnight, and found the place plundered, the door and window open, and the fugitives gone. There was no letter or message, nor has he heard a word since. He at once gave the alarm to the police.'

Holmes brooded for some minutes.

'You say he was painting. What was he painting?'

'Well, he was painting the passage. But he had already painted the door and woodwork of this room I spoke of.'

'Does it strike you as a strange occupation in the circumstances?'

' "One must do something to ease an aching heart." That was his own explanation. It was eccentric, no doubt, but he is clearly an eccentric man. He tore up one of his wife's photographs in my presence—tore it up furiously in a tempest of passion. "I never wish to see her damned face again," he shrieked.'

'Anything more, Watson?'

'Yes, one thing which struck me more than anything else. I had driven to the Blackheath Station and had caught my train there, when just as it was starting I saw a man dart into the carriage next to my own. You know that I have a quick eye for faces, Holmes. It was undoubtedly the tall, dark man whom I had addressed in the street. I saw him once more at London Bridge,* and then I lost him in the crowd. But I am convinced that he was following me.'

'No doubt! No doubt!' said Holmes. 'A tall, dark, heavily-moustached man, you say, with grey-tinted sunglasses?'

'Holmes, you are a wizard. I did not say so, but he *had* grey-tinted sun-glasses.'

'And a Masonic tie-pin?'*

'Holmes!'

'Quite simple, my dear Watson. But let us get down to what is practical. I must admit to you that the case, which seemed to me to be so absurdly simple as to be hardly worth my notice, is rapidly assuming a very different aspect. It is true that though in your mission you have missed everything of importance, yet even those things which have obtruded themselves upon your notice give rise to serious thought.'

'What have I missed?'

'Don't be hurt, my dear fellow. You know that I am quite impersonal. No one else would have done better. Some possibly not so well. But clearly you have missed some vital points. What is the opinion of the neighbours about this man Amberley and his wife? That surely is of importance. What of Dr Ernest? Was he the gay Lothario* one would expect? With your natural advantages, Watson, every lady is your helper and accomplice. What about the girl at the post office, or the wife of the greengrocer? I can picture you whispering soft nothings with the young lady at the "Blue Anchor", and receiving hard somethings in exchange. All this you have left undone.'

'It can still be done.'

'It has been done. Thanks to the telephone* and the help of the Yard, I can usually get my essentials without leaving this room. As a matter of fact, my information confirms the man's story. He has the local repute of being a miser as well as a harsh and exacting husband. That he had a large sum of money in that strong-room of his is certain. So also is it that young Dr Ernest, an unmarried man, played chess with Amberley, and probably played the fool with his wife. All this seems plain sailing, and one would think that there was no more to be said—and yet!—and yet!'

'Where lies the difficulty?'

'In my imagination, perhaps. Well, leave it there, Watson. Let us escape from this weary workaday world by the side door of music. Carina sings to-night at the Albert Hall,* and we still have time to dress, dine and enjoy.'

In the morning I was up betimes,* but some toast crumbs and two empty eggshells told me that my companion was earlier still. I found a scribbled note on the table.

Dear Watson,

There are one or two points of contact which I should wish to establish with Mr Josiah Amberley. When I have done so we can dismiss the case—or not. I would only ask you to be on hand about three o'clock, as I conceive it possible that I may want you.

S. H.

I saw nothing of Holmes all day, but at the hour named he returned, grave, preoccupied and aloof. At such times it was easier to leave him to himself.

'Has Amberley been here yet?'

'No.'

'Ah! I am expecting him.'

He was not disappointed, for presently the old fellow arrived with a very worried and puzzled expression upon his austere face.

'I've had a telegram, Mr Holmes. I can make nothing of it.' He handed it over, and Holmes read it aloud.

Come at once without fail. Can give you information as to your recent loss.—ELMAN. The Vicarage.

'Dispatched at two-ten from Little Purlington,'* said Holmes. 'Little Purlington is in Essex, I believe, not far from Frinton.* Well, of course you will start at once. This is evidently from a responsible person, the vicar of the place. Where is my Crockford?* Yes, here we have him. J. C. Elman, MA, Living* of Mossmoor cum* Little Purlington. Look up the trains, Watson.'

'There is one at five-twenty from Liverpool Street.'*

'Excellent. You had best go with him, Watson. He may need help or advice. Clearly we have come to a crisis in this affair.'

But our client seemed by no means eager to start.

'It's perfectly absurd, Mr Holmes,' he said. 'What can this man possibly know of what has occurred? It is waste of time and money.'

'He would not have telegraphed to you if he did not know something. Wire* at once that you are coming.'

'I don't think I shall go.'

Holmes assumed his sternest aspect.

'It would make the worst possible impression both on the police and upon myself, Mr Amberley, if when so obvious a clue arose you should refuse to follow it up. We should feel that you were not really in earnest in this investigation.'

Our client seemed horrified at the suggestion.

'Why, of course I shall go if you look at it in that way,' said he. 'On the face of it, it seems absurd to suppose that this parson knows anything, but if you think—'

'I *do* think,' said Holmes, with emphasis, and so we were launched upon our journey. Holmes took me aside before we left the room and gave me one word of counsel which showed that he considered the matter to be of importance. 'Whatever you do, see that he really *does* go,' said he. 'Should he break away or return, get to the nearest telephone exchange and send the single word "Bolted". I will arrange here that it shall reach me wherever I am.'

Little Purlington is not an easy place to reach, for it is on a branch line. My remembrance of the journey is not a pleasant one, for the weather was hot, the train slow, and my companion sullen and silent, hardly talking at all, save to make an occasional sardonic remark as to the futility of our proceedings. When we at last reached the little station it was a two-mile drive before we came to the Vicarage, where a big, solemn, rather pompous clergyman received us in his study. Our telegram lay before him.

'Well, gentlemen,' he asked, 'what can I do for you?'

'We came,' I explained, 'in answer to your wire.'

'My wire! I sent no wire.'

'I mean the wire which you sent to Mr Josiah Amberley about his wife and his money.'

'If this is a joke, sir, it is a very questionable one,' said the vicar angrily. 'I have never heard of the gentleman you name, and I have not sent a wire to anyone.'

Our client and I looked at each other in amazement.

'Perhaps there is some mistake,' said I; 'are there perhaps two vicarages? Here is the wire itself, signed Elman, and dated from the Vicarage.'

'There is only one vicarage, sir, and only one vicar, and this wire is a scandalous forgery, the origin of which shall certainly be investigated by the police. Meanwhile, I can see no possible object in prolonging this interview.'

So Mr Amberley and I found ourselves on the roadside in what seemed to me to be the most primitive village in England. We made for the telegraph office,* but it was already closed. There was a telephone, however, at the little 'Railway Arms', and by it I got into touch with Holmes, who shared in our amazement at the result of our journey.

'Most singular!' said the distant voice. 'Most remarkable! I much fear, my dear Watson, that there is no return train tonight. I have unwittingly condemned you to the horrors of a country inn. However, there is always Nature, Watson—

Nature and Josiah Amberley—you can be in close commune with both.' I heard his dry chuckle as he turned away.

It was soon apparent to me that my companion's reputation as a miser was not undeserved. He had grumbled at the expense of the journey, had insisted upon travelling third-class, and was now clamorous in his objections to the hotel bill. Next morning, when we did at last arrive in London, it was hard to say which of us was in the worse humour.

'You had best take Baker Street as we pass,' said I. 'Mr Holmes may have some fresh instructions.'

'If they are not worth more than the last ones they are not of much use,' said Amberley, with a malevolent scowl. None the less, he kept me company. I had already warned Holmes by telegram of the hour of our arrival, but we found a message waiting that he was at Lewisham, and would expect us there. That was a surprise, but an even greater one was to find that he was not alone in the sitting-room of our client. A stern-looking, impassive man sat beside him, a dark man with grey-tinted glasses and a large Masonic pin projecting from his tie.

'This is my friend Mr Barker,' said Holmes. 'He has been interesting himself also in your business, Mr Josiah Amberley, though we have been working independently. But we have both the same question to ask you!'

Mr Amberley sat down heavily. He sensed impending danger. I read it in his straining eyes and his twitching features.

'What is the question, Mr Holmes?'

'Only this: What did you do with the bodies?'

The man sprang to his feet with a hoarse scream. He clawed into the air with his bony hands. His mouth was open and for the instant he looked like some horrible bird of prey. In a flash we got a glimpse of the real Josiah Amberley, a misshapen demon with a soul as distorted as his body. As he fell back into his chair he clapped his hand to his lips as if to stifle a cough. Holmes sprang at his throat like a tiger, and twisted his face towards the ground. A white pellet* fell from between his gasping lips.

'No short cuts, Josiah Amberley. Things must be done decently and in order.* What about it, Barker?'

'I have a cab at the door,' said our taciturn companion.

'It is only a few hundred yards to the station. We will go together. You can stay here, Watson. I shall be back within half an hour.'

The old colourman had the strength of a lion in that great trunk of his, but he was helpless in the hands of the two experienced man-handlers. Wriggling and twisting, he was dragged to the waiting cab, and I was left to my solitary vigil in the ill-omened house. In less time than he had named, however, Holmes was back, in company with a smart young police inspector.

'I've left Barker to look after the formalities,' said Holmes. 'You had not met Barker, Watson. He is my hated rival upon the Surrey shore. When you said a tall dark man it was not difficult for me to complete the picture. He has several good cases to his credit, has he not, Inspector?'

'He has certainly interfered several times,' the Inspector answered with reserve.

'His methods are irregular, no doubt, like my own. The irregulars are useful sometimes, you know. You, for example, with your compulsory warning about whatever he said being used against him, could never have bluffed this rascal into what is virtually a confession.'

'Perhaps not. But we get there all the same, Mr Holmes. Don't imagine that we had not formed our own views of this case, and that we would not have laid our hands on our man. You will excuse us for feeling sore when you jump in with methods which we cannot use, and so rob us of the credit.'

'There shall be no such robbery, MacKinnon. I assure you that I efface myself from now onwards, and as to Barker, he had done nothing save what I told him.'

The Inspector seemed considerably relieved.

'That is very handsome of you, Mr Holmes. Praise or blame can matter little to you, but it is very different to us when the newspapers begin to ask questions.'

'Quite so. But they are pretty sure to ask questions anyhow, so it would be as well to have answers. What will you say, for example, when the intelligent and enterprising reporter asks you what the exact points were which aroused your suspicion, and finally gave you a certain conviction as to the real facts?'

The Inspector looked puzzled.

'We don't seem to have got any real facts yet, Mr Holmes. You say that the prisoner, in the presence of three witnesses, practically confessed, by trying to commit suicide, that he had murdered his wife and her lover. What other facts have you?'

'Have you arranged for a search?'

'There are three constables on their way.'

'Then you will soon get the clearest fact of all. The bodies cannot be far away. Try the cellars and the garden. It should not take long to dig up the likely places. This house is older than the water-pipes. There must be a disused well somewhere. Try your luck there.'

'But how did you know of it, and how was it done?'

'I'll show you first how it was done, and then I will give the explanation which is due to you, and even more to my long-suffering friend here, who has been invaluable throughout. But, first, I would give you an insight into this man's mentality. It is a very unusual one—so much so that I think his destination is more likely to be Broadmoor* than the scaffold. He has, to a high degree, the sort of mind which one associates with the medieval Italian nature rather than with the modern Briton. He was a miserable miser who made his wife so wretched by his niggardly ways that she was a ready prey for any adventurer. Such a one came upon the scene in the person of this chess-playing doctor. Amberley excelled at chess—one mark, Watson, of a scheming mind. Like all misers, he was a jealous man, and his jealousy became a frantic mania. Rightly or wrongly, he suspected an intrigue. He determined to have his revenge, and he planned it with diabolical cleverness. Come here!'

Holmes led us along the passage with as much certainty as if he had lived in the house, and halted at the open door of the strong-room.

'Pooh! What an awful smell of paint!' cried the Inspector.

'That was our first clue,' said Holmes. 'You can thank Dr Watson's observation for that, though he failed to draw the inference. It set my foot upon the trail. Why should this man at such a time be filling his house with strong odours? Obviously, to cover some other smell which he wished to conceal—some guilty smell which would suggest suspicions. Then came the idea of a room such as you see here with the iron door and shutter—a hermetically sealed* room. Put those two facts together, and whither do they lead? I could only determine that by examining the house myself. I was already certain that the case was serious, for I had examined the box-office chart at the Haymarket Theatre—another of Dr Watson's bull's-eyes—and ascertained that neither B thirty nor thirty-two of the upper circle had been occupied that night. Therefore, Amberley had not been to the theatre, and his *alibi* fell to the ground. He made a bad slip when he allowed my astute friend to notice the number of the seat taken for his wife. The question now arose how I might be able to examine the house. I sent an agent to the most impossible village I could think of, and summoned my man to it at such an hour that he could not possibly get back. To prevent any miscarriage, Dr Watson accompanied him. The good vicar's name I took, of course, out of my Crockford. Do I make it all clear to you?'

'It is masterly,' said the Inspector, in an awed voice.

'There being no fear of interruption I proceeded to burgle the house. Burglary has always been an alternative profession, had I cared to adopt it, and I have little doubt that I should have come to the front.* Observe what I found. You see the gas-pipe along the skirting here. Very good. It rises in the angle of the wall, and there is a tap here in the corner. The pipe runs out into the strong-room, as you can see, and ends in that plastered rose* in the centre of the ceiling, where it is concealed by the ornamentation. That end is wide open. At any moment by turning the outside tap the room could be flooded with gas. With door and shutter closed and the tap full on I would not give two minutes of conscious sensation to anyone shut up in that little chamber. By what devilish device he decoyed them there I do not know, but once inside the door they were at his mercy.'

The Inspector examined the pipe with interest. 'One of our officers mentioned the smell of gas,' said he, 'but, of course, the window and door were open then, and the paint—or some of it—was already about. He had begun the work of painting the day before, according to his story. But what next, Mr Holmes?'

'Well, then came an incident which was rather unexpected to myself. I was slipping through the pantry window in the early dawn when I felt a hand inside my collar, and a voice said: "Now, you rascal, what are you doing in there?" When I could twist my head round I looked into the tinted spectacles of my friend and rival, Mr Barker. It was curious foregathering,* and set us both smiling. It seems that he had been engaged by Dr Ray Ernest's family to make some investigations, and had come to the same conclusion as to foul play. He had watched the house for some days, and had spotted Dr Watson as one of the obviously suspicious characters who had called there. He could hardly arrest Watson, but when he saw a man actually climbing out of the pantry window there came a limit to his restraint. Of course, I told him how matters stood and we continued the case together.'

'Why him? Why not us?'

'Because it was in my mind to put that little test which answered so admirably. I fear you would not have gone so far.'

The Inspector smiled.

'Well, maybe not. I understand that I have your word, Mr Holmes, that you step right out of the case now and that you turn all your results over to us.'

'Certainly, that is always my custom.'

'Well, in the name of the Force I thank you. It seems a clear case, as you put it, and there can't be much difficulty over the bodies.'

'I'll show you a grim little bit of evidence,' said Holmes, 'and I am sure Amberley himself never observed it. You'll get results, Inspector, by always putting yourself in the other fellow's place, and thinking what you would do yourself. It takes some imagination, but it pays. Now, we will suppose that you were shut up in this little room, had not two minutes to live, but wanted to get even with the fiend who was probably mocking at you from the other side of the door. What would you do?'

'Write a message.'

'Exactly. You would like to tell people how you died. No use writing on paper. That would be seen. If you wrote on the wall some eye might rest upon it. Now, look here! Just above the skirting is scribbled with a purple indelible pencil: "We we—" That's all.'

'What do you make of that?'

'Well, it's only a foot above the ground. The poor devil was on the floor and dying when he wrote it. He lost his senses before he could finish.'

'He was writing, "We were murdered." '

'That's how I read it. If you find an indelible pencil on the body—'*

'We'll look out for it, you may be sure. But those securities? Clearly there was no robbery at all. And yet he *did* possess those bonds. We verified that.'

'You may be sure he has them hidden in a safe place. When the whole elopement had passed into history he would suddenly discover them, and announce that the guilty couple had relented and sent back the plunder or had dropped it on the way.'

'You certainly seem to have met every difficulty,' said the Inspector. 'Of course, he was bound to call us in, but why he should have gone to you I can't understand.'

'Pure swank!'* Holmes answered. 'He felt so clever and so sure of himself that he imagined no one could touch him. He could say to any suspicious neighbour, "Look at the steps I have taken. I have consulted not only the police, but even Sherlock Holmes." '

The Inspector laughed.

'We must forgive you your "even", Mr Holmes,' said he; 'it's as workmanlike a job as I can remember.'

A couple of days later my friend tossed across to me a copy of the bi-weekly *North Surrey Observer*.* Under a series of flaming headlines, which began with 'The Haven Horror' and ended with 'Brilliant Police Investigation', there was a packed column of print which gave the first consecutive account of the affair. The concluding paragraph is typical of the whole. It ran thus:

> The remarkable acumen by which Inspector MacKinnon deduced from the smell of paint that some other smell, that of gas, for example, might be concealed; the bold deduction that the strong-room might also be the death-chamber, and the subsequent inquiry which led to the discovery of the bodies in a disused well, cleverly concealed by a dog-kennel, should live in the history of crime as a standing example of the intelligence of our professional detectives.

'Well, well, MacKinnon is a good fellow,' said Holmes, with a tolerant smile. 'You can file it in our archives, Watson. Some day the true story may be told.'*

The Veiled Lodger

WHEN one considers that Mr Sherlock Holmes was in active practice for twenty-three years, and that during seventeen of these I was allowed to co-operate with him and to keep notes of his doings, it will be clear that I have a mass of material at my command. The problem has always been, not to find, but to choose. There is the long row of year-books which fill a shelf, and there are the dispatch-cases filled with documents, a perfect quarry for the student not only of crime, but of the social and official scandals* of the late Victorian era. Concerning these latter, I may say that the writers of agonized letters, who beg that the honour of their families or the reputation of famous forebears may not be touched, have nothing to fear. The discretion and high sense of professional honour which have always distinguished my friend are still at work in the choice of these memoirs, and no confidence will be abused. I deprecate, however, in the strongest way the attempts which have been made lately to get at and to destroy these papers. The source of these outrages is known, and if they are repeated I have Mr Holmes's authority for saying that the whole story concerning the politician, the lighthouse, and the trained cormorant* will be given to the public. There is at least one reader who will understand.

It is not reasonable to suppose that every one of these cases gave Holmes the opportunity of showing those curious gifts of instinct and observation which I have endeavoured to set forth in these memoirs. Sometimes he had with much effort to pick the fruit, sometimes it fell easily into his lap. But the most terrible human tragedies were often involved in these cases which brought him the fewest personal opportunities, and it is one of these which I now desire to record. In telling it, I have made a slight change of name and place, but otherwise the facts are as stated.

One forenoon*—it was late in 1896—I received a hurried note from Holmes asking for my attendance. When I arrived, I found him seated in a smoke-laden atmosphere, with an elderly, motherly woman of the buxon landlady type in the corresponding chair in front of him.

'This is Mrs Merrilow, of South Brixton,'* said my friend, with a wave of the hand. 'Mrs Merrilow does not object to tobacco, Watson, if you wish to indulge your filthy habits. Mrs Merrilow has an interesting story to tell which may well lead to further developments in which your presence may be useful.'

'Anything I can do—'

'You will understand, Mrs Merrilow, that if I come to Mrs Ronder I should prefer to have a witness. You will make her understand that before we arrive.'

'Lord bless you, Mr Holmes,' said our visitor, 'she is that anxious to see you that you might bring the whole parish at your heels!'

'Then we shall come early in the afternoon. Let us see that we have our facts correct before we start. If we go over them it will help Dr Watson to understand the situation. You say that Mrs Ronder has been your lodger for seven years and that you have only once seen her face.'

'And I wish to God I had not!' said Mrs Merrilow.

'It was, I understand, terribly mutilated.'

'Well, Mr Holmes, you would hardly say it was a face at all. That's how it looked. Our milkman got a glimpse of her once peeping out of the upper window, and he dropped his tin* and the milk all over the front garden. That is the kind of face it is. When I saw her—I happened on her unawares—she covered up quick, and then she said, "Now, Mrs Merrilow, you know at last why it is that I never raise my veil." '

'Do you know anything about her history?'

'Nothing at all.'

'Did she give references when she came?'

'No, sir, but she gave hard cash, and plenty of it. A quarter's rent right down on the table in advance and no arguing about terms. In these times a poor woman like me can't afford to turn down a chance like that.'

'Did she give any reason for choosing your house?'

'Mine stands well back from the road and is more private than most. Then again, I only take the one, and I have no family of my own. I reckon she had tried others and found that mine suited her best. It's privacy she is after, and she is ready to pay for it.'

'You say that she never showed her face from first to last save on the one accidental occasion. Well, it is a very remarkable story, most remarkable, and I don't wonder that you want it examined.'

'I don't, Mr Holmes, I am quite satisfied so long as I get my rent. You could not have a quieter lodger, or one who gives less trouble.'

'Then what has brought matters to a head?'

'Her health, Mr Holmes. She seems to be wasting away. And there's something terrible on her mind. "Murder!" she cries. "Murder!" And once I heard her, "You cruel beast! You monster!" she cried. It was in the night, and it fair rang through the house and sent the shivers through me. So I went to her in the morning. "Mrs Ronder," I says, "if you have anything that is troubling your soul, there's the clergy," I says, "and there's the police. Between them you should get some help." "For God's sake, not the police!" says she, "and the clergy can't change what is past. And yet," she says, "it would ease my mind if someone knew the truth before I died." "Well," says I, "if you won't have the regulars, there is this detective man what we read about"—beggin' your pardon, Mr Holmes. And she, she fair jumped at it. "That's the man," says she. "I wonder I never thought of it before. Bring him here, Mrs Merrilow, and if he won't come, tell him I am the wife of Ronder's wild-beast show. Say that, and give him the name Abbas Parva."* Here it is as she wrote it, Abbas Parva. "That will bring him, if he's the man I think he is." '

'And it will, too,' remarked Holmes. 'Very good, Mrs Merrilow. I should like to have a little chat with Dr Watson.

That will carry us till lunch-time. About three o'clock you may expect to see us at your house in Brixton.'

Our visitor had no sooner waddled out of the room—no other verb can describe Mrs Merrilow's method of progression—than Sherlock Holmes threw himself with fierce energy upon the pile of commonplace books in the corner. For a few minutes there was a constant swish of the leaves, and then with a grunt of satisfaction he came upon what he sought. So excited was he that he did not rise, but sat upon the floor like some strange Buddha,* with crossed legs, the huge books all round him, and one open upon his knees.

'The case worried me at the time, Watson. Here are my marginal notes to prove it. I confess that I could make nothing of it. And yet I was convinced that the coroner was wrong. Have you no recollection of the Abbas Parva tragedy?'

'None, Holmes.'

'And yet you were with me then. But certainly my own impression was very superficial, for there was nothing to go by, and none of the parties had engaged my services. Perhaps you would care to read the papers?'

'Could you not give me the points?'

'That is very easily done. It will probably come back to your memory as I talk. Ronder, of course, was a household word. He was the rival of Wombwell, and of Sanger,* one of the greatest showmen of his day. There is evidence, however, that he took to drink, and that both he and his show were on the down grade at the time of the great tragedy. The caravan had halted for the night at Abbas Parva, which is a small village in Berkshire, when this horror occurred. They were on their way to Wimbledon, travelling by road, and they were simply camping, and not exhibiting, as the place is so small a one that it would not have paid them to open.

'They had among their exhibits a very fine North African lion. Sahara King* was its name, and it was the habit, both of Ronder and his wife, to give exhibitions inside its cage. Here, you see, is a photograph of the performance, by which you will perceive that Ronder was a huge porcine* person, and that his wife was a very magnificent woman. It was deposed at the inquest that there had been some signs that the lion was dangerous, but, as usual, familiarity begat contempt, and no notice was taken of the fact.

'It was usual for either Ronder or his wife to feed the lion at night. Sometimes one went, sometimes both, but they never allowed anyone else to do it, for they believed that so long as they were the food-carriers he would regard them as benefactors, and would never molest them. On this particular night, seven years ago, they both went, and a very terrible happening followed, the details of which have never been made clear.

'It seems that the whole camp was roused near midnight by the roars of the animal and the screams of the woman. The different grooms and *employés* rushed from their tents, carrying lanterns, and by their light an awful sight was revealed. Ronder lay, with the back of his head crushed in and deep claw-marks across his scalp, some ten yards from the cage, which was open. Close to the door of the cage lay Mrs Ronder, upon her back, with the creature squatting and snarling above her. It had torn her face in such a fashion that it was never thought that she could live. Several of the circus men, headed by Leonardo,* the strong man, and Griggs,* the clown, drove the creature off with poles, upon which it sprang back into the cage, and was at once locked in. How it had got loose was a mystery. It was conjectured that the pair intended to enter the cage, but that when the door was loosed the creature bounded out upon them. There was no other point of interest in the evidence, save that the woman in a delirium of agony kept screaming, "Coward! Coward!" as she was carried back to the van* in which they lived. It was six months before she was fit to give evidence, but the inquest was duly held with the obvious verdict of death from misadventure.'

'What alternative could be conceived?' said I.

'You may well say so. And yet there were one or two points which worried young Edmunds, of the Berkshire Constabulary. A smart lad that! He was sent later to Allahabad.* That was how I came into the matter, for he dropped in and smoked a pipe or two over it.'

'A thin, yellow-haired man?'

'Exactly. I was sure you would pick up the trail presently.'

'But what worried him?'

'Well, we were both worried. It was so deucedly difficult to reconstruct the affair. Look at it from the lion's point of view. He is liberated. What does he do? He takes half a dozen bounds forward, which brings him to Ronder. Ronder turns to fly—the claw-marks were on the back of his head—but the lion strikes him down. Then, instead of bounding on and escaping, he returns to the woman, who was close to the cage, and he knocks her over and chews her face up. Then, again, those cries of hers would seem to imply that her husband had in some way failed her. What could the poor devil have done to help her? You see the difficulty?'

'Quite.'

'And then there was another thing. It comes back to me now as I think it over. There was some evidence that, just at the time the lion roared and the woman screamed, a man began shouting in terror.'

'This man Ronder, no doubt.'

'Well, if his skull was smashed in you would hardly expect to hear from him again. There were at least two witnesses who spoke of the cries of a man being mingled with those of a woman.'

'I should think the whole camp was crying out by then. As to the other points, I think I could suggest a solution.'

'I should be glad to consider it.'

'The two were together, ten yards from the cage, when the lion got loose. The man turned and was struck down. The woman conceived the idea of getting into the cage and shutting the door.* It was her only refuge. She made for it, and just as she reached it the beast bounded after her and knocked her over. She was angry with her husband for having encouraged the beast's rage by turning. If they had faced it, they might have cowed it. Hence her cries of "Coward!" '

'Brilliant, Watson! Only one flaw in your diamond.'

'What is the flaw, Holmes?'

'If they were both ten paces from the cage, how came the beast to get loose?'

'Is it possible that they had some enemy who loosed it?'

'And why should it attack them savagely when it was in the habit of playing with them, and doing tricks with them inside the cage?'

'Possibly the same enemy had done something to enrage it.'

Holmes looked thoughtful and remained in silence for some moments.

'Well, Watson, there is this to be said for your theory. Ronder was a man of many enemies. Edmunds told me that in his cups he was horrible. A huge bully of a man, he cursed and slashed at everyone who came in his way. I expect those cries about a monster, of which our visitor has spoken, were nocturnal reminiscences of the dear departed.* However, our speculations are futile until we have all the facts. There is a cold partridge on the sideboard, Watson, and a bottle of Montrachet.* Let us renew our energies before we make a fresh call upon them.'

When our hansom deposited us at the house of Mrs Merrilow, we found that plump lady blocking up the open door of her humble but retired abode. It was very clear that her chief preoccupation was lest she should lose a valuable lodger, and she implored us, before showing us up, to say and do nothing which could lead to so undesirable an end. Then, having reassured her, we followed her up the straight badly-carpeted staircase and were shown into the room of the mysterious lodger.

It was a close, musty, ill-ventilated place, as might be expected, since its inmate seldom left it. From keeping beasts in a cage, the woman seemed, by some retribution of Fate, to have become herself a beast in a cage. She sat now in a broken arm-chair in the shadowy corner of the room. Long years of inaction had coarsened the lines of her figure, but at some period it must have been beautiful, and was still full and voluptuous. A thick dark veil covered her face, but it was cut off close at her upper lip, and disclosed a perfectly-shaped mouth and a delicately-rounded chin. I could well conceive that she had indeed been a very remarkable woman. Her voice, too, was well-modulated and pleasing.

'My name is not unfamiliar to you, Mr Holmes,' said she. 'I thought that it would bring you.'

'That is so, madam,* though I do not know how you are aware that I was interested in your case.'

'I learned it when I had recovered my health and was examined by Mr Edmunds, the County detective. I fear I lied to him. Perhaps it would have been wiser had I told the truth.'

'It is usually wiser to tell the truth. But why did you lie to him?'

'Because the fate of someone else depended upon it. I know that he was a very worthless being, and yet I would not have his destruction upon my conscience. We had been so close—so close!'

'But has this impediment been removed?'

'Yes, sir. The person that I allude to is dead.'

'Then why should you not now tell the police anything you know?'

'Because there is another person to be considered. That other person is myself. I could not stand the scandal and publicity which would come from a police examination. I have not long to live, but I wish to die undisturbed. And yet I wanted to find one man of judgement to whom I could tell my terrible story, so that when I am gone all might be understood.'

'You compliment me, madam. At the same time, I am a responsible person. I do not promise you that when you have spoken I may not myself think it my duty to refer the case to the police.'

'I think not, Mr Holmes. I know your character and methods too well, for I have followed your work for some years. Reading is the only pleasure which Fate has left me, and I miss little which passes in the world. But in any case, I will take my chance of the use which you may make of my tragedy. It will ease my mind to tell it.'

'My friend and I would be glad to hear it.'

The woman rose and took from a drawer the photograph of a man. He was clearly a professional acrobat, a man of magnificent physique, taken with his huge arms folded across his swollen chest and a smile breaking from under his heavy moustache—the self-satisfied smile of the man of many conquests.

'That is Leonardo,' she said.

'Leonardo, the strong man, who gave evidence?'

'The same. And this—this is my husband.'

It was a dreadful face—a human pig, or rather a human wild boar, for it was formidable in its bestiality. One could imagine that vile mouth champing and foaming in its rage, and one could conceive those small, vicious eyes darting pure malignancy as they looked forth upon the world. Ruffian, bully, beast—it was all written on that heavy-jowled face.

'Those two pictures will help you, gentlemen, to understand the story. I was a poor circus girl brought up on the sawdust, and doing springs through the hoop before I was ten. When I became a woman this man loved me, if such lust as his can be called love, and in an evil moment I became his wife. From that day I was in hell, and he the devil who tormented me. There was no one in the show who did not know of this treatment. He deserted me for others. He tied me down and lashed me with his riding-whip when I complained. They all pitied me and they all loathed him, but what could they do? They feared him, one and all. For he was terrible at all times, and murderous when he was drunk. Again and again he was had for assault, and for cruelty to the beasts, but he had plenty of money, and the fines were nothing to him. The best men all left us, and the show began to go downhill. It was only Leonardo and I who kept it up—with little Jimmy Griggs, the clown. Poor devil, he had not much to be funny about, but he did what he could to hold things together.

'Then Leonardo came more and more into my life. You see what he was like. I know now the poor spirit that was hidden in that splendid body, but compared to my husband he seemed like the Angel Gabriel.* He pitied me and helped me, till at last our intimacy turned to love—deep, deep, passionate love, such love as I had dreamed of, but never hoped to feel. My husband suspected it, but I think that he was a coward as well as a bully, and that Leonardo was the one man that he was afraid of. He took revenge in his own way by torturing me more than ever. One night my cries brought Leonardo to the door of our van. We were near tragedy that night, and soon my lover and I understood that it could not be avoided. My husband was not fit to live. We planned that he should die.

'Leonardo had a clever, scheming brain. It was he who planned it. I do not say that to blame him, for I was ready to go with him every inch of the way. But I should never have had the wit to think of such a plan. We made a club—Leonardo made it—and in the leaden head he fastened five long steel nails, the points outwards, with just such a spread as the lion's paw. This was to give my husband his death-blow, and yet to leave the evidence that it was the lion which we would loose who had done the deed.

'It was a pitch-dark night when my husband and I went down, as was our custom, to feed the beast. We carried with us the raw meat in a zinc* pail. Leonardo was waiting at the corner of the big van which we should have to pass before we reached the cage. He was too slow, and we walked past him before he could strike, but he followed us on tiptoe and I heard the crash as the club smashed my husband's skull. My heart leaped with joy at the sound. I sprang forward, and I undid the catch which held the door of the great lion's cage.

'And then the terrible thing happened. You may have heard how quick these creatures are to scent human blood, and how it excites them. Some strange instinct had told the creature in one instant that a human being had been slain. As I slipped the bars it bounded out, and was on me in an instant. Leonardo could have saved me. If he had rushed forward and struck the beast with his club he might have cowed it. But the man lost his nerve. I heard him shout in his terror, and then I saw him turn and fly. At the same instant the teeth of the lion met in my face. Its hot, filthy breath had already poisoned me and I was hardly conscious of pain. With the palms of my hands I tried to push the great steaming, blood-stained jaws away from me, and I screamed for help. I was conscious that the camp was stirring, and then dimly I remember a group of men, Leonardo, Griggs, and others, dragging me from under the creature's paws. That was my last memory, Mr Holmes, for many a weary month. When I came to myself, and saw myself in the mirror, I cursed that lion—oh, how I cursed him!—not because he had torn away my beauty, but because he had not torn away my life. I had but one desire, Mr Holmes, and I had enough money to gratify it. It was that I should cover myself so that my poor face should be seen by none, and that I should dwell where none whom I had ever known should find me. That was all that was left to me to do—and that is what I have done. A poor wounded beast that has crawled into its hole to die—that is the end of Eugenia Ronder.'

We sat in silence for some time after the unhappy woman had told her story. Then Holmes stretched out his long arm and patted her hand with such a show of sympathy as I had seldom known him to exhibit.

'Poor girl!' he said. 'Poor girl! The ways of Fate are indeed hard to understand. If there is not some compensation hereafter, then the world is a cruel jest. But what of this man Leonardo?'

'I never saw him or heard from him again. Perhaps I have been wrong to feel so bitterly against him. He might as soon have loved one of the freaks whom we carried round the country as the thing which the lion had left. But a woman's love is not so easily set aside. He had left me under the beast's claws, he had deserted me in my need, and yet I could not bring myself to give him to the gallows. For myself, I cared nothing what became of me. What could be more dreadful than my actual life? But I stood between Leonardo and his fate.'

'And he is dead?'

'He was drowned last month when bathing near Margate.* I saw his death in the paper.'

'And what did he do with this five-clawed club, which is the most singular and ingenious part of all your story?'

'I cannot tell, Mr Holmes. There is a chalk-pit by the camp, with a deep green pool at the base of it. Perhaps in the depths of that pool—'

'Well, well, it is of little consequence now. The case is closed.'

'Yes,' said the woman, 'the case is closed.'

We had risen to go, but there was something in the woman's voice which arrested Holmes's attention. He turned swiftly upon her.

'Your life is not your own,' he said. 'Keep your hands off it.'

'What use is it to anyone?'

'How can you tell? The example of patient suffering is in itself the most precious of all lessons to an impatient world.'

The woman's answer was a terrible one. She raised her veil and stepped forward into the light.

'I wonder if you would bear it,' she said.

It was horrible. No words can describe the framework of a face when the face itself is gone. Two living and beautiful brown eyes looking sadly out from that grisly ruin did but make the view more awful. Holmes held up his hand in a gesture of pity and protest, and together we left the room.

Two days later, when I called upon my friend, he pointed with some pride to a small blue bottle upon his mantelpiece. I picked it up. There was* a red poison label. A pleasant almondy odour rose when I opened it.

'Prussic acid?'* said I.

'Exactly. It came by post. "I send you my temptation. I will follow your advice." That was the message. I think, Watson, we can guess the name of the brave woman who sent it.'

*Shoscombe Old Place**

SHERLOCK HOLMES had been bending for a long time over a low-power microscope. Now he straightened himself up and looked round at me in triumph.

'It is glue,* Watson,' said he. 'Unquestionably it is glue. Have a look at these scattered objects in the field!'*

I stooped to the eyepiece and focused for my vision.

'Those hairs are threads from a tweed coat. The irregular grey masses are dust. There are epithelial scales* on the left. Those brown blobs in the centre are undoubtedly glue.'

'Well,' I said, laughing, 'I am prepared to take your word for it. Does anything depend upon it?'

'It is a very fine demonstration,' he answered. 'In the St Pancras* case you may remember that a cap was found beside the dead policeman. The accused man denies that it is his. But he is a picture-frame maker who habitually handles glue.'

'Is it one of your cases?'

'No; my friend, Merivale of the Yard, asked me to look into the case. Since I ran down that coiner by the zinc and copper filings* in the seam of his cuff they have begun to realize the importance of the microscope.'* He looked impatiently at his watch. 'I had a new client calling, but he is overdue. By the way, Watson, you know something of racing?'

'I ought to. I pay for it with about half my wound pension.'*

'Then I'll make you my "Handy Guide to the Turf".* What about Sir Robert Norberton? Does the name recall anything?'

'Well, I should say so.* He lives at Shoscombe Old Place, and I know it well, for my summer quarters* were down there once. Norberton nearly came within your province once.'

'How was that?'

'It was when he horsewhipped Sam Brewer, the well-known Curzon Street* moneylender, on Newmarket Heath.* He nearly killed the man.'

'Ah, he sounds interesting! Does he often indulge in that way?'

'Well, he has the name of being a dangerous man. He is about the most daredevil rider in England—second in the Grand National a few years back. He is one of those men who have overshot their true generation. He should have been a buck* in the days of the Regency*—a boxer, an athlete, a plunger on the Turf,* a lover of fair ladies, and, by all account, so far down Queer Street* that he may never find his way back again.'

'Capital, Watson. A thumb-nail sketch. I seem to know the man. Now, can you give me some idea of Shoscombe Old Place?'

'Only that it is in the centre of Shoscombe Park, and that the famous Shoscombe stud* and training quarters are to be found there.'

'And the head trainer', said Holmes, 'is John Mason. You need not look surprised at my knowledge, Watson, for this is a letter from him which I am unfolding. But let us have some more about Shoscombe. I seem to have struck a rich vein.'

'There are the Shoscombe spaniels,' said I. 'You hear of them at every dog show. The most exclusive breed in England. They are the special pride of the lady of Shoscombe Old Place.'

'Sir Robert Norberton's wife, I presume!'

'Sir Robert has never married. Just as well, I think, considering his prospects. He lives with his widowed sister, Lady Beatrice Falder.'

'You mean that she lives with him?'

'No, no. The place belonged to her late husband, Sir James. Norberton has no claim on it at all. It is only a life interest and reverts to her husband's brother. Meantime, she draws the rents every year.'

'And brother Robert, I suppose, spends the said rents?'

'That is about the size of it. He is a devil of a fellow and must lead her a most uneasy life. Yet I have heard that she is devoted to him. But what is amiss at Shoscombe?'

'Ah, that is just what I want to know. And here, I expect, is the man who can tell us.'

The door opened and the page had shown in a tall, clean-shaven man with the firm, austere expression which is only seen upon those who have to control horses or boys. Mr John Mason had many of both under his sway, and he looked equal to the task. He bowed with cold self-possession and seated himself upon the chair to which Holmes had waved him.

'You had my note, Mr Holmes?'

'Yes, but it explained nothing.'

'It was too delicate a thing for me to put the details on paper. And too complicated. It was only face to face I could do it.'

'Well, we are at your disposal.'

'First of all, Mr Holmes, I think that my employer, Sir Robert, has gone mad.'

Holmes raised his eyebrows. 'This is Baker Street, not Harley Street,'* said he. 'But why do you say so?'

'Well, sir, when a man does one queer thing, or two queer things, there may be a meaning to it, but when everything he does is queer, then you begin to wonder. I believe Shoscombe Prince and the Derby* have turned his brain.'

'That is a colt you are running?'

'The best in England, Mr Holmes. I should know, if anyone does. Now, I'll be plain with you, for I know you are gentlemen of honour and that it won't go beyond the room. Sir Robert has got to win this Derby. He's up to the neck,* and it's his last chance. Everything he could raise or borrow is on the horse—and at fine odds, too! You can get forties now, but it was nearer the hundred when he began to back him.'

'But how is that, if the horse is so good?'

'The public don't know how good he is. Sir Robert has been too clever for the touts.* He has the Prince's half-brother out for spins.* You can't tell 'em apart. But there are two lengths in a furlong between them when it comes to a gallop. He thinks of nothing but the horse and the race. His whole life is on it. He's holding off the Jews* till then. If the Prince fails him, he is done.'

'It seems a rather desperate gamble, but where does the madness come in?'

'Well, first of all, you have only to look at him. I don't believe he sleeps at night. He is down at the stables at all hours. His eyes are wild. It has all been too much for his nerves. Then there is his conduct to Lady Beatrice!'

'Ah! what is that?'

'They have always been the best of friends. They had the same tastes, the two of them, and she loved the horses as much as he did. Every day at the same hour she would drive down to see them—and, above all she loved the Prince. He would prick up his ears when he heard the wheels on the gravel, and he would trot out each morning to the carriage to get his lump of sugar. But that's all over now.'

'Why?'

'Well, she seems to have lost all interest in the horses. For a week now she has driven past the stables with never so much as "good morning"!'

'You think there has been a quarrel?'

'And a bitter, savage, spiteful quarrel at that. Why else would he give away her pet spaniel that she loved as if he were her child? He gave it a few days ago to old Barnes, what keeps the "Green Dragon", three miles off, at Crendall.'*

'That certainly did seem strange.'

'Of course, with her weak heart and dropsy* one couldn't expect that she could get about with him, but he spent two hours every evening in her room. He might well do what he could, for she has been a rare good friend to him. But that's all over, too. He never goes near her. And she takes it to heart. She is brooding and sulky and drinking, Mr Holmes—drinking like a fish.'

'Did she drink before this estrangement?'

'Well, she took her glass, but now it is often a whole bottle of an evening. So Stephens, the butler, told me. It's all changed, Mr Holmes, and there is something damned rotten about it. But then, again, what is master doing down at the old church crypt at night? And who is the man that meets him there?'

Holmes rubbed his hands.

'Go on, Mr Mason. You get more and more interesting.'

'It was the butler who saw him go. Twelve o'clock at night and raining hard. So next night I was up at the house and, sure enough, master was off again. Stephens and I went after him, but it was jumpy work, for it would have been a bad job if he had seen us. He's a terrible man with his fists if he gets started, and no respecter of persons. So we were shy of getting too near, but we marked him down all right. It was the haunted crypt that he was making for, and there was a man waiting for him there.'

'What is this haunted crypt?'

'Well, sir, there is an old ruined chapel in the park. It is so old that nobody could fix its date. And under it there's a crypt which has a bad name among us. It's a dark, damp, lonely place by day, but there are few in that county that would have the nerve to go near it at night. But master's not afraid. He never feared anything in his life. But what is he doing there in the night-time?'

'Wait a bit!' said Holmes. 'You say there is another man there. It must be one of your own stable-men, or someone from the house! Surely you have only to spot who it is and question him?'

'It's no one I know.'

'How can you say that?'

'Because I have seen him, Mr Holmes. It was on that second night. Sir Robert turned and passed us—me and Stephens, quaking in the bushes like two bunny-rabbits, for there was a bit of moon that night. But we could hear the other moving about behind. We were not afraid of him. So we up when Sir Robert was gone and pretended we were just having a walk like in the moonlight, and so we came right on him as casual and innocent as you please. "Hullo, mate! who may you be?" says I. I guess he had not heard us coming, so he looked over his shoulder with a face as if he had seen the Devil coming out of Hell. He let out a yell, and away he went as hard as he could lick it in the darkness. He could run!—I'll give him that. In a minute he was out of sight and hearing, and who he was, or what he was, we never found.'

'But you saw him clearly in the moonlight?'

'Yes, I would swear to his yellow face—a mean dog, I should say. What could he have in common with Sir Robert?'

Holmes sat for some time lost in thought.

'Who keeps Lady Beatrice Falder company?' he asked at last.

'There is her maid, Carrie Evans. She has been with her this five years.'

'And is, no doubt, devoted?'

Mr Mason shuffled uncomfortably.

'She's devoted enough,' he answered at last. 'But I won't say to whom.'

'Ah!' said Holmes.

'I can't tell tales out of school.'

'I quite understand, Mr Mason. Of course, the situation is clear enough. From Dr Watson's description of Sir Robert I can realize that no woman is safe from him. Don't you think the quarrel between brother and sister may lie there?'

'Well, the scandal has been pretty clear for a long time.'

'But she may not have seen it before. Let us suppose that she has suddenly found it out. She wants to get rid of the woman. Her brother will not permit it. The invalid, with her weak heart and inability to get about, has no means of enforcing her will. The hated maid is still tied to her. The lady refuses to speak, sulks, takes to drink. Sir Robert in his anger takes her pet spaniel away from her. Does not all this hang together?'

'Well, it might do—so far as it goes.'

'Exactly! As far as it goes. How would all that bear upon the visits by night to the old crypt? We can't fit that into our plot.'

'No, sir, and there's something more that I can't fit in. Why should Sir Robert want to dig up a dead body?'

Holmes sat up abruptly.

'We only found it out yesterday—after I had written to you. Yesterday Sir Robert had gone to London, so Stephens and I went down to the crypt. It was all in order, sir, except that in one corner was a bit of a human body.'

'You informed the police, I suppose?'

Our visitor smiled grimly.

'Well, sir, I think it would hardly interest them. It was just the head and a few bones of a mummy. It may have been a thousand years old. But it wasn't there before. That I'll swear, and so will Stephens. It had been stowed away in a corner and covered over with a board, but that corner had always been empty before.'

'What did you do with it?'

'Well, we just left it there.'

'That was wise. You say Sir Robert was away yesterday. Has he returned?'

'We expect him back to-day.'

'When did Sir Robert give away his sister's dog?'

'It was just a week ago to-day. The creature was howling outside the old well-house, and Sir Robert was in one of his tantrums that morning. He caught it up and I thought he would have killed it. Then he gave it to Sandy Bain, the jockey, and told him to take the dog to old Barnes at the "Green Dragon", for he never wished to see it again.'

Holmes sat for some time in silent thought. He had lit the oldest and foulest of his pipes.

'I am not clear yet what you want me to do in this matter, Mr Mason,' he said at last. 'Can't you make it more definite?'

'Perhaps this will make it more definite, Mr Holmes,' said our visitor.

He took a paper from his pocket and, unwrapping it carefully, he exposed a charred fragment of bone.

Holmes examined it with interest.

'Where did you get it?'

'There is a central heating furnace in the cellar under Lady Beatrice's room. It's been off for some time, but Sir Robert complained of cold and had it on again. Harvey runs it—he's one of my lads. This very morning he came to me with this which he found raking out the cinders. He didn't like the look of it.'

'Nor do I,' said Holmes. 'What do you make of it, Watson?'

It was burned to a black cinder, but there could be no question as to its anatomical significance.

'It's the upper condyle* of a human femur,' said I.

'Exactly!' Holmes had become very serious. 'When does this lad tend* to the furnace?'

'He makes it up every evening and then leaves it.'

'Then anyone could visit it during the night?'

'Yes, sir.'

'Can you enter it from outside?'

'There is one door from outside. There is another which leads up by a stair to the passage in which Lady Beatrice's room is situated.'

'These are deep waters, Mr Mason; deep and rather dirty. You say that Sir Robert was not at home last night?'

'No, sir.'

'Then, whoever was burning bones, it was not he.'

'That's true, sir.'

'What is the name of that inn you spoke of?'

'The "Green Dragon".'

'Is there good fishing in that part of Berkshire?'

The honest trainer showed very clearly upon his face that he was convinced that yet another lunatic had come into his harassed life.

'Well, sir, I've heard there are trout in the mill-stream and pike in the Hall lake.'

'That's good enough. Watson and I are famous fishermen—are we not, Watson? You may address us in future at the "Green Dragon". We should reach it to-night. I need not say that we don't want to see you, Mr Mason, but a note will reach us, and no doubt I could find you if I want you. When we have gone a little farther into the matter I will let you have a considered opinion.'

Thus it was that on a bright May evening Holmes and I found ourselves alone in a first-class carriage and bound for the little 'halt-on-demand' station* of Shoscombe. The rack above us was covered with a formidable litter of rods, reels and baskets. On reaching our destination a short drive took us to an old-fashioned tavern, where a sporting host, Josiah Barnes,* entered eagerly into our plans for the extirpation of the fish of the neighbourhood.

'What about the Hall lake and the chance of a pike?' said Holmes.

The face of the innkeeper clouded.

'That wouldn't do, sir. You might chance to find yourself in the lake before you were through.'

'How's that, then?'

'It's Sir Robert, sir. He's terrible jealous of touts. If you two strangers were as near his training quarters as that he'd be after you as sure as fate. He ain't taking no chances, Sir Robert ain't.'

'I've heard he has a horse entered for the Derby.'

'Yes, and a good colt, too. He carries all our money for the race, and all Sir Robert's into the bargain. By the way'—he looked at us with thoughtful eyes—'I suppose you ain't on the Turf yourselves?'

'No, indeed. Just two weary Londoners who badly need some good Berkshire air.'

'Well, you are in the right place for that. There is a deal of it lying about. But mind what I have told you about Sir Robert. He's the sort that strikes first and speaks afterwards. Keep clear of the park.'

'Surely, Mr Barnes! We certainly shall. By the way, that was a most beautiful spaniel that was whining in the hall.'

'I should say it was. That was the real Shoscombe breed. There ain't a better in England.'

'I am a dog-fancier myself,' said Holmes. 'Now, if it is a fair question, what would a prize dog like that cost?'

'More than I could pay, sir. It was Sir Robert himself who gave me this one. That's why I have to keep it on a lead. It would be off to the Hall in a jiffy if I gave it its head.'

'We are getting some cards in our hand, Watson,' said Holmes, when the landlord had left us. 'It's not an easy one to play, but we may see our way in a day or two. By the way, Sir Robert is still in London, I hear. We might, perhaps, enter the sacred domain to-night without fear of bodily assault. There are one or two points on which I should like reassurance.'

'Have you any theory, Holmes?'

'Only this, Watson, that *something* happened a week or so ago which has cut deep into the life of the Shoscombe household. What is that something? We can only guess at it from its effects. They seem to be of a curiously mixed character. But that should surely help us. It is only the colourless, uneventful case which is hopeless.

'Let us consider our data. The brother no longer visits the beloved invalid sister. He gives away her favourite dog. Her *dog*, Watson! Does that suggest nothing to you?'

'Nothing but the brother's spite.'

'Well, it might be so. Or—well, there is an alternative. Now, to continue our review of the situation from the time that the quarrel, if there is a quarrel, began. The lady keeps her room, alters her habits, is not seen save when she drives out with her maid, refuses to stop at the stables to greet her favourite horse, and apparently takes to drink. That covers the case, does it not?'

'Save for the business in the crypt.'

'That is another line of thought. There are two, and I beg you will not tangle them. Line A, which concerns Lady Beatrice, has a vaguely sinister flavour, has it not?'

'I can make nothing of it.'

'Well, now, let us take up line B, which concerns Sir Robert. He is mad keen upon winning the Derby. He is in the hands of the Jews, and may at any moment be sold up and his racing stables seized by his creditors. He is a daring and desperate man. He derives his income from his sister. His sister's maid is his willing tool. So far we seem to be on fairly safe ground, do we not?'

'But the crypt?'

'Ah, yes, the crypt! Let us suppose, Watson—it is merely a scandalous supposition, a hypothesis put forward for argument's sake—that Sir Robert has done away with his sister.'

'My dear Holmes, it is out of the question.'*

'Very possibly, Watson. Sir Robert is a man of an honourable stock. But you do occasionally find a carrion crow among the eagles.* Let us for a moment argue upon this supposition. He could not fly the country until he had realized his fortune, and that fortune could only be realized by bringing off this coup with Shoscombe Prince. Therefore he has still to stand his ground. To do this he would have to dispose of the body of his victim, and he would also have to find a substitute who would impersonate her. With the maid as his confidante that would not be impossible. The woman's body might be conveyed to the crypt, which is a place so seldom visited, and it might be secretly destroyed at night in the furnace, leaving behind it such evidence as we have already seen. What say you to that, Watson?'

'Well, it is all possible if you grant the original monstrous supposition.'

'I think that there is a small experiment which we may try to-morrow, Watson, in order to throw some light on the matter. Meanwhile, if we mean to keep up our characters, I suggest that we have our host in for a glass of his own wine and hold some high converse upon eels and dace,* which seems to be the straight road to his affections. We may chance to come upon some useful local gossip in the process.'

In the morning Holmes discovered that we had come without our spoon-bait* for jack,* which absolved us from fishing for the day. About eleven o'clock we started for a walk, and he obtained leave to take the black spaniel with us.

'This is the place,' said he, as we came to two high park gates with heraldic griffins* towering above them. 'About midday, Mr Barnes informs me, the old lady takes a drive, and the carriage must slow down while the gates are opened. When it comes through, and before it gathers speed, I want you, Watson, to stop the coachman with some question. Never mind me. I shall stand behind this holly-bush and see what I can see.'

It was not a long vigil. Within a quarter of an hour we saw the big open yellow barouche* coming down the long avenue, with two splendid, high-stepping grey carriage horses in the shafts. Holmes crouched behind his bush with the dog. I stood unconcernedly swinging a cane in the roadway. A keeper ran out and the gates swung open.

The carriage had slowed to a walk and I was able to get a good look at the occupants. A highly-coloured young woman with flaxen hair and impudent eyes sat on the left. At her right was an elderly person with rounded back and a huddle of shawls about her face and shoulders which proclaimed the invalid. When the horses reached the high road I held up my hand with an authoritative gesture, and as the coachman pulled up I inquired if Sir Robert was at Shoscombe Old Place.

At the same moment Holmes stepped out and released the spaniel. With a joyous cry it dashed forward to the carriage and sprang upon the step. Then in a moment its eager greeting changed to furious rage, and it snapped at the back skirt above it.

'Drive on! Drive on!' shrieked a harsh voice. The coachman lashed the horses, and we were left standing in the roadway.

'Well, Watson, that's done it,' said Holmes, as he fastened the lead to the neck of the excited spaniel. 'He thought it was his mistress and he found it was a stranger. Dogs don't make mistakes.'*

'But it was the voice of a man!' I cried.

'Exactly! We have added one card to our hand, Watson, but it needs careful playing, all the same.'

My companion seemed to have no further plans for the day, and we did actually use our fishing tackle in the mill-stream, with the result that we had a dish of trout for our supper. It was only after that meal that Holmes showed signs of renewed activity. Once more we found ourselves upon the same road as in the morning, which led us to the park gates. A tall, dark figure was awaiting us there, who proved to be our London acquaintance, Mr John Mason, the trainer.

'Good evening, gentlemen,' said he. 'I got your note, Mr Holmes. Sir Robert has not returned yet, but I hear that he is expected to-night.'

'How far is this crypt from the house?' asked Holmes.

'A good quarter of a mile.'

'Then I think we can disregard him altogether.'

'I can't afford to do that, Mr Holmes. The moment he arrives he will want to see me to get the last news of Shoscombe Prince.'

'I see! In that case we must work without you, Mr Mason. You can show us the crypt and then leave us.'

It was pitch-dark and without a moon, but Mason led us over the grass-lands until a dark mass loomed up in front of us which proved to be the ancient chapel. We entered the broken gap which was once the porch, and our guide, stumbling among heaps of loose masonry, picked his way to the corner of the building, where a steep stair led down into the crypt. Striking a match, he illuminated the melancholy place—dismal and evil-smelling, with ancient crumbling walls of rough-hewn stone, and piles of coffins, some of lead and some of stone, extending upon one side right up to the arched and groined* roof which lost itself in the shadows above our heads. Holmes had lit his lantern which shot a tiny tunnel of vivid yellow light upon the mournful scene. Its rays were reflected back from the coffin-plates, many of them adorned with the griffin and coronet of this old family which carried its honours even to the gate of Death.

'You spoke of some bones, Mr Mason. Could you show them before you go?'

'They are here in this corner.' The trainer strode across and then stood in silent surprise as our light was turned upon the place. 'They are gone,' said he.

'So I expected,' said Holmes, chuckling. 'I fancy the ashes of them might even now be found in that oven which had already consumed a part.'

'But why in the world would anyone want to burn the bones of a man who has been dead a thousand years?' asked John Mason.

'That is what we are here to find out,' said Holmes. 'It may mean a long search, and we need not detain you. I fancy that we shall get our solution before morning.'

When John Mason had left us, Holmes set to work making a very careful examination of the graves, ranging from a very ancient one, which appeared to be Saxon,* in the centre, through a long line of Norman Hugos and Odos,* until we reached the Sir William and Sir Denis Falder of the eighteenth century. It was an hour or more before Holmes came to a leaden coffin standing on end before the entrance to the vault. I heard his little cry of satisfaction, and was aware from his hurried but purposeful movements that he had reached a goal. With his lens he was eagerly examining the edges of the heavy lid. Then he drew from his pocket a short jemmy,* a box-opener, which he thrust into a chink, levering back the whole front, which seemed to be secured by only a couple of clamps. There was a rending, tearing sound as it gave way, but it had hardly hinged back and partly revealed the contents before we had an unforeseen interruption.

Someone was walking in the chapel above. It was the firm, rapid step of one who came with a definite purpose and knew well the ground upon which he walked. A light streamed down the stairs, and an instant later the man who bore it was framed in the Gothic archway. He was a terrible figure, huge in stature and fierce in manner. A large stable-lantern which he held in front of him shone upwards upon a strong, heavily-moustached face and angry eyes, which glared round him into every recess of the vault, finally fixing themselves with a deadly stare upon my companion and myself.

'Who the devil are you?' he thundered. 'And what are you doing upon my property?' Then, as Holmes returned no answer, he took a couple of steps forward and raised a heavy stick which he carried. 'Do you hear me?' he cried. 'Who are you? What are you doing here?' His cudgel quivered in the air.

But instead of shrinking, Holmes advanced to meet him.

'I also have a question to ask you, Sir Robert,' he said in his sternest tone. 'Who is this? And what is it doing here?'

He turned and tore open the coffin-lid behind him. In the glare of the lantern I saw a body swathed in a sheet from head to foot, with dreadful, witch-like features, all nose and chin, projecting at one end, the dim, glazed eyes staring from a discoloured and crumbling face.

The Baronet* had staggered back with a cry and supported himself against a stone sarcophagus.*

'How came you to know of this?' he cried. And then, with some return of his truculent manner: 'What business is it of yours?'

'My name is Sherlock Holmes,' said my companion. 'Possibly it is familiar to you. In any case, my business is that of every other good citizen—to uphold the law. It seems to me that you have much to answer for.'

Sir Robert glared for a moment, but Holmes's quiet voice and cool, assured manner had their effect.

' 'Fore God, Mr Holmes, it's all right,' said he. 'Appearances are against me, I'll admit, but I could act no otherwise.'*

'I should be happy to think so, but I fear your explanations must be for the police.'

Sir Robert shrugged his broad shoulders.

'Well, if it must be, it must. Come up to the house and you can judge for yourself how the matter stands.'

A quarter of an hour later we found ourselves in what I judge, from the lines of polished barrels behind glass covers, to be the gun-room of the old house. It was comfortably furnished, and here Sir Robert left us for a few moments. When he returned he had two companions with him; the one, the florid young woman whom we had seen in the carriage; the other, a small rat-faced man with a disagreeably furtive manner. These two wore an appearance of utter bewilderment, which showed that the Baronet had not yet had time to explain to them the turn events had taken.

'There', said Sir Robert, with a wave of his hand, 'are Mr and Mrs Norlett. Mrs Norlett, under her maiden name of Evans, has for some years been my sister's confidential maid. I have brought them here because I feel that my best course is to explain the true position to you, and they are the two people upon earth who can substantiate what I say.'

'Is this necessary, Sir Robert? Have you thought what you are doing?' cried the woman.

'As to me, I entirely disclaim all responsibility,' said her husband.

Sir Robert gave him a glance of contempt. 'I will take all responsibility,' said he. 'Now, Mr Holmes, listen to a plain statement of the facts.

'You have clearly gone pretty deeply into my affairs or I should not have found you where I did. Therefore, you know already, in all probability, that I am running a dark horse* for the Derby and that everything depends upon my success. If I win, all is easy. If I lose—well, I dare not think of that!'

'I understand the position,' said Holmes.

'I am dependent upon my sister, Lady Beatrice, for everything. But it is well known that her interest in the estate is for her own life only. For myself, I am deeply in the hands of the Jews. I have always known that if my sister were to die my creditors would be on to my estate* like a flock of vultures. Everything would be seized; my stables, my horses—everything. Well, Mr Holmes, my sister *did* die just a week ago.'

'And you told no one!'

'What could I do? Absolute ruin faced me. If I could stave things off for three weeks all would be well. Her maid's husband—this man here—is an actor. It came into our heads—it came into my head—that he could for that short period personate my sister. It was but a case of appearing daily in the carriage, for no one need enter her room save the maid. It was not difficult to arrange. My sister died of the dropsy which had long afflicted her.'

'That will be for a coroner to decide.'

'Her doctor would certify that for months her symptoms have threatened such an end.'

'Well, what did you do?'

'The body could not remain there. On the first night Norlett and I carried it out to the old well-house, which is now never used. We were followed, however, by her pet spaniel, which yapped continually at the door so I felt some safer place was needed. I got rid of the spaniel and we carried the body to the crypt of the church. There was no indignity or irreverence, Mr Holmes. I do not feel that I have wronged the dead.'

'Your conduct seems to me inexcusable, Sir Robert.'

The Baronet shook his head impatiently 'It is easy to preach,' said he. 'Perhaps you would have felt differently if you had been in my position. One cannot see all one's hopes and all one's plans shattered at the last moment and make no effort to save them. It seemed to me that it would be no unworthy resting-place if we put her for the time in one of the coffins of her husband's ancestors lying in what is still consecrated ground. We opened such a coffin, removed the contents, and placed her as you have seen her. As to the old relics which we took out, we could not leave them on the floor of the crypt. Norlett and I removed them, and he descended at night and burned them in the central furnace. There is my story, Mr Holmes, though how you forced my hand so that I have to tell it is more than I can say.'

Holmes sat for some time lost in thought.

'There is one flaw in your narrative, Sir Robert,' he said at last. 'Your bets on the race, and therefore your hopes for the future, would hold good even if your creditors seized your estate.'

'The horse would be part of the estate. What do they care for my bets? As likely as not they would not run him at all. My chief creditor is, unhappily, my most bitter enemy—a rascally fellow, Sam Brewer, whom I was once compelled to horsewhip on Newmarket Heath. Do you suppose that he would try to save me?'

'Well, Sir Robert,' said Holmes, rising, 'this matter must, of course, be referred to the police. It was my duty to bring the facts to light and there I must leave it. As to the morality or decency of your own conduct, it is not for me to express an opinion. It is nearly midnight, Watson, and I think we may make our way back to our humble abode.'*

It is generally known now that this singular episode ended upon a happier note than Sir Robert's actions deserved. Shoscombe Prince did win the Derby, the sporting owner did net eighty thousand pounds in bets, and the creditors did hold their hand until the race was over, when they were paid in full, and enough was left to re-establish Sir Robert in a fair position in life. Both police and coroner took a lenient view of the transaction, and beyond a mild censure for the delay in registering the lady's decease, the lucky owner got away scatheless* from this strange incident in a career which has now outlived its shadows and promises to end* in an honoured old age.*

EXPLANATORY NOTES

The first English edition of *The Case-Book of Sherlock Holmes*, containing 12 stories, was published by John Murray on 16 June 1927, in an edition of 15,150 copies. The Colonial issue (5000 copies), in Murray's Imperial Library, and the first American edition (by the George H. Doran Company of New York) were published on the same day. All the stories had first appeared in the *Strand Magazine*, from Oct. 1921 to Apr. 1927. This edition follows the order of first publication (which, as far as can be determined, reflects the order of composition). When published in book form by Murray the stories were arranged as follows:

'The Adventure of the Illustrious Client'
'The Adventure of the Blanched Soldier'
'The Adventure of the Mazarin Stone'
'The Adventure of the Three Gables'
'The Adventure of the Sussex Vampire'
'The Adventure of the Three Garridebs'
'The Problem of Thor Bridge'
'The Adventure of the Creeping Man'
'The Adventure of the Lion's Mane'
'The Adventure of the Veiled Lodger'
'The Adventure of Shoscombe Old Place'
'The Adventure of the Retired Colourman'

THE MAZARIN STONE

First published in the *Strand Magazine*, 62 (Oct. 1921), 288–98, with 3 illustrations by A. Gilbert. First American publication in *Hearst's International Magazine*, New York, 40 (Nov. 1921), 6–8, 64–5, with 4 illustrations by Frederick Dorr Steele.

5 *Billy*: Master Charles Chaplin (1889–1977) played this part in William Gillette's *Sherlock Holmes* from 27 July 1903 until 5 Mar. 1906.

saturnine: sombre.

said of him: a counterpart to the last paragraph of 'His Last Bow', where Holmes salutes Watson as 'the one fixed point in a changing age'.

6 *Lord Cantlemere*: evidently one of the royal circle, possibly a government-appointed member of the royal household, or else with official status conferred by the monarch. Perhaps the status—but not necessarily the character—of someone like Reginald Baliol Brett, second Viscount Esher (1852–1930), a liaison figure between Edward VII and cabinet ministers, was in ACD's mind. Traditionally, royal circles included self-important drones, of which ACD may have seen specimens during the arrangements for his own knighthood in 1902 (see his comments on palace protocol in *Memories and Adventures* (1930)).

facsimile: an exact copy.

'We used something of the sort once before': a reference to 'The Empty House' (*Return*).

8 *Count Negretto Sylvius*: literally, 'little black wood-dweller', from *negretto* (Italian) and *silva* (the Latin for wood); hence 'Count Black Wood'; presumably a private joke at the expense of *Blackwood's Magazine*, long and unavailingly courted by ACD in the 1880s until it published 'A Physiologist's Wife' (Sept. 1890)—though only after he demanded its withdrawal when it remained unused for a year after acceptance. He never wrote for *Blackwood's* again, to the magazine's regret. In the original version of 'The Mazarin Stone', a one-act play called *The Crown Diamond: or, an Evening with Sherlock Holmes*, the Sylvius character is Colonel Sebastian Moran, retaining from 'The Empty House' (*Return*) his big-game huntsmanship and being deceived by a bust of Holmes. *The Crown Diamond* opened at the Bristol Hippodrome on 2 May 1921 and toured for eighteen months. (For the text, see Richard Lancelyn Green, *The Uncollected Sherlock Holmes*.)

Crown Jewel: 'jewel' in all but late editions, but it is clearly intended as one of the official regalia (which accounts for the involvement of Cantlemere, the Premier, and the Home Secretary in their official capacities). It is not meant as a jewel belonging to the monarch (like the emerald tie-pin Queen Victoria gives Holmes in 'The Bruce-Partington Plans'). Presumably, the Mazarin Stone became a British Crown Diamond as a result of the French Revolution, by gift or purchase from exiled French royalty or aristocracy.

the great yellow Mazarin Stone: presumably named after Cardinal Jules Mazarin (1602–61), Italian-born statesman who ruled France during the minority of Louis XIV.

8 *Sam Merton*: the oafish and criminal pugilist would appear to have been given a name deliberately suggestive of *Sandford and Merton* (three volumes, 1783, 1787, 1789), the famous edifying children's tale by Thomas Day (1748–89). It was frequently cited in ACD's day, with several modern parodies.

gudgeon: a small, easily caught, carp-like freshwater fish; a traditional Scots symbol of stupidity.

9 *Straubenzee's*: fictitious. Major-General Sir Casimir Cartwright Van Straubenzee (1867–1956) was Inspector-General of the Royal Artillery 1917–18.

the Minories: a street in the City of London, which takes its name from the community of the *Sorores Minores* (Franciscan nuns or 'Poor Clares'), established in 1293.

'In the waiting room, sir': a hitherto unmentioned feature of 221B Baker Street.

10 *Youghal of the CID*: this is the only mention of Youghal in the series. The name is that of the easternmost Cork port, on the Blackwater, near the property of ACD's Foley cousins of west Waterford. *CID*: Criminal Investigation Department; its London headquarters was Scotland Yard.

11 *Tavernier*: fictitious. It means a publican or inn-keeper.

13 *in this book*: this links 'The Mazarin Stone' with the later story 'The Illustrious Client' (as well as with the much earlier 'The Empty House'), the book there being Baron Gruner's own work. In that story, Sir James Damery is an effective reworking of the Cantlemere character, whilst Baron Gruner improves on Count Sylvius as the voluptuary/predator/criminal figure. 'The Mazarin Stone', while a weak story in itself, is thus of interest as showing the process of creative development from 'The Empty House', through the play *The Crown Diamond* and 'The Mazarin Stone' itself, to 'The Illustrious Client', and can be seen as a transition piece, rather than a true Holmes story.

train-de-luxe: French name for a train made up exclusively of first-class coaches.

Riviera: stretch of the French and Italian coastlines between the Alps and the Mediterranean; a fashionable resort for the wealthy.

Credit Lyonnais: one of the principal French banks.

14 *Whitehall*: a spacious street linking Trafalgar Square with Westminster and containing many government offices. A Crown Jewel would normally be in the Tower of London; presumably, it was temporarily in a government department while being cleaned, lent (back to the French perhaps?), or as part of an exhibition.

Commissionaire: a member of the Corps of Commissionaires (established in 1859), a uniformed body of retired soldiers. As well as being found on duty outside government buildings they would be expected to carry messages, go on errands, and act as escorts for visitors. Several commissionaires figure in the Holmes stories, for example in *A Study in Scarlet* and 'The Blue Carbuncle' (*Adventures*).

peached: informed (cf. Peachem in John Gay's *The Beggar's Opera* [1728]).

compound a felony: the offence Holmes will commit is, rather, misprision of felony. He is a private person and is concealing knowledge of a criminal offence. Were he an official he could 'compound', i.e. agree for a consideration not to prosecute.

15 *prize-fighter*: a professional boxer.

16 *the Hoffmann Barcarolle*: the Barcarolle (a boatman's song) by the French composer Jacques Offenbach (1819–80), from his *Tales of Hoffmann* (1881).

split: informed.

'I'll do him down a thick 'un': the meaning is approximately 'I'll give him a good thrashing.' To 'do down' someone means to get the better of them. A 'thick 'un' was a sovereign, a coin worth a pound sterling and distinguished by its thickness.

swing: hang.

leary: sometimes spelt 'leery': clever, sly.

17 *Madame Tussaud*: famous waxworks in the Marylebone Road in London. It was set up in 1802 by Madame Marie Grosholtz Tussaud (1760–1850). Before moving to its present site in 1884 it was in Baker Street.

quid: pound sterling; plural identical with singular. Possibly originating in the Latin for 'what'.

17 *better men than he*: adapted from the last line of Rudyard Kipling's poem 'Gunga Din', an appropriate literary source for a big-game hunter, however criminal.

Amsterdam: largest city of the Netherlands, particularly noted for its diamond market.

Lime Street: on the south side of Leadenhall Street, so called after the lime-burners.

19 *a second door*: another surprise feature of 221B Baker Street.

These modern gramophones: gramophones, unlike phonographs, used disc recordings rather than cylinders. Guy Warrack (*Sherlock Holmes and Music*, 1947) thought that a recording for unaccompanied violin of the Hoffmann Barcarolle is unlikely; but Holmes might have made a private recording of the piece himself. John Robert Moore (*Modern Languages Quarterly*, 8/1, Mar. 1947) pointed out the parallel between the use of the gramophone in 'The Mazarin Stone' and in 'The Japanned Box' (*Round the Fire Stories*, 1897). Also relevant is ACD's use of it in 'The Voice of Science', the first story of his to be published in the newly founded *Strand* (Apr. 1891).

mid-Victorian whiskers: a moustache and sideboards in the style fashionable in the 1860s.

20 *glossy blackness*: presumably dyed, like the hair of Colonel Cochrane in ACD's novel *The Tragedy of the Korosko* (1896).

receiver: a person who has accepted stolen property, usually by paying considerably less than the item's value.

THOR BRIDGE

First published in the *Strand Magazine*, 63 (Feb./Mar. 1922), 95–104, 211–17, with 3 illustrations by A. Gilbert. First American publication in *Hearst's International Magazine*, New York, 41 (Feb./Mar. 1922), 6–7, 69; 14–15, 60–2, with 3 illustrations by G. Patrick Nelson.

23 *Cox and Co.*: destroyed in the Blitz of the Second World War.

Charing Cross: the area of London incorporating Trafalgar Square, the northern end of Whitehall, and the western part of the Strand.

Indian Army: strictly speaking, Watson was never in the Indian Army. The regiments with which he served were part of the regular establishment of the British Army and were merely ordered to India; the Indian army was entirely separate.

Mr James Phillimore: perhaps a version of the vanishing of Benjamin Bathurst (b. 1784), secretary of the British Legation at Leghorn, who in 1809, while *en route* to England, walked round to the back of a coach in Vienna and was never seen again.

the cutter Alicia: a cutter is a small, one-masted sailing vessel; cf. the brigantine *Mary Celeste*, found abandoned, for no known reason, in a high-running sea between the Azores and Portugal in 1872. See also ACD's early story 'J. Habakkuk Jephson's Statement' (1884).

Isadora Persano: his first name is feminine, which has led some commentators to conjecture that he might have been a transvestite.

unfathomed cases: an echo perhaps of Thomas Gray's 'Elegy in a Country Churchyard' (1751), 'Full many a gem of purest ray serene, / The dark unfathom'd caves of ocean bear'.

24 *told as by a third person*: of the Sherlock Holmes short stories (apart from the two narrated by Holmes himself), only two ('His Last Bow' and 'The Mazarin Stone') are not narrated by Watson. *A Study in Scarlet* and *The Valley of Fear* also have third-person flashbacks.

my own experience: a conscious distancing from the narrative voice of 'The Mazarin Stone' (see previous note).

Family Herald: a popular magazine for women first published in 1842. It ceased publication in 1940. It specialized in romantic melodramas and is satirized in P. G. Wodehouse's *If I Were You* (1931).

for: Anglification of American usage 'from'. A British Member of Parliament sits on behalf of a constituency, but the federal structure of the USA requires that a Representative or Senator be a citizen of the State in question—hence 'from' it.

gold-mining magnate: only California (Alaska did not become a State until 1959) had sufficient gold to make a Gold King's constituency. The most conspicuous case of a Senator from California becoming a millionaire through mine ownership and speculation was George Hearst (1820–91), appointed 1886, elected 1888; he died in office. The public and private lives of his more famous son, the newspaper magnate William Randolph Hearst (1863–1951), seem remarkably reminiscent in style to those of J. Neil Gibson. If this is indeed a portrait of the younger Hearst, though using the political status and economic origins of the older, it invites comparison with Orson Welles's film *Citizen Kane* (1941). The American publication of 'Thor Bridge' in *Hearst's International Magazine* may suggest that ACD gleaned information about Hearst's character and methods from his employees (cf. Marlow Bates): his work had been appearing in Hearst's magazine since 1919, and he must have detested Hearst's Anglophobia, then at its height.

25 *Hampshire*: a southern English county much loved by ACD.

tragic end: the correct expression, as opposed to the vulgarly tautologous 'tragic death' (which implies distinction from a plenitude of comic deaths).

coroner's jury: a coroner's chief function was to preside over inquiries into the causes of accidental or suspicious deaths; his jury consisted of twelve men. The office does not exist in ACD's native Scotland.

Assizes: a court held from time to time in all the counties of England and Wales, presided over by a judge, at which those indicted with serious crimes were tried by jury.

Winchester: ancient cathedral city in Hampshire.

Miss Dunbar: a Scottish name, recalling the Scots poet William Dunbar (*c.* 1460–1530), whose famous line *Timor Mortis conturbat me* (the fear of death shakes me [*Lament for the Makeris*, ?1507]) may have prompted ACD's use of 'Mortimer' to denote a herald of death in *The Hound of the Baskervilles* and as an originator of a peculiarly fearful death in 'The Devil's Foot' (*His Last Bow*).

26 *calibre*: the size of the bore of a gun's barrel.

two juries: at the coroner's court and the police court.

27 *balustraded sides*: with short pillars, linked at the top by a rail or coping.

reed-girt: amid reed-beds.

Mere: lake or pool, with the implication of murky water.

Billy: see the openings of 'The Mazarin Stone' and *The Valley of Fear*.

28 *Abraham Lincoln*: sixteenth President of the USA (born 1809, assassinated 1865) who maintained the Union throughout the Civil War. He grew his beard during the war, at a little girl's suggestion, to humanize his appearance.

29 *every paper*: as close as the story comes to suggesting that Gibson is a Hearstian magnate and manipulator of news. Holmes's reply acknowledges the hollowness of the 'booming' process (Hearst was even more famous for its opposite).

booming you: i.e. greatly enlarging your reputation, with the consequent opportunity of financial gain.

the talk of two continents: why only two?

32 *Manaos*: chief city of northern Brazil.

33 *Amazon*: the great river of South America, vividly described in ACD's *The Lost World* (1912).

possession: the sexual vocabulary of capitalism.

34 *ruin*: a characteristic Holmes re-reading, in this instance asserting the real meaning of Gibson's 'possession'.

molested: in this context, sexually harassed.

35 *the latter*: Mrs Gibson's having no cause for 'body-jealousy' is of course contradicted by the earlier admission of molestation, but it is a shrewdly observed instance of male efforts to maintain a façade while knowing that it is already crumbling.

permits: required in order to interview Grace Dunbar in Winchester prison.

36 *wonderful fine*: while using American, Irish, and Scots common speech, ACD avoided dialect, save in such usages as this.

out of the road: out of the way (usage more characteristic of Scotland than England).

37 *pheasant preserves*: the part of a country estate in which pheasant and other game birds are reared for shooting.

fruitful line of enquiry: the first instalment in the *Strand* broke off here.

38 *reserved for the Assizes*: Grace Dunbar would not give her answer to the charges made against her until her trial at the Assize court—a customary but not invariable defence procedure.

41 *rising barrister*: a barrister is a lawyer qualified to represent the interests of his client in a higher court of law. It is surprising (or perhaps not surprising) that Gibson does not retain the services of a senior barrister (a Queen's Counsel).

42 *After seeing you*: contrast Holmes's guardedness about Miss Morstan (the future Mrs Watson) in *The Sign of the Four.*

45 *the God of justice*: American editions have 'god'—perhaps reflecting the separation of Church and State in the USA.

46 *safety-catch*: a small catch which, when fastened, prevents a gun being fired accidently.

trap: a two-wheeled, open horse-carriage.

49 *earthly lessons are taught*: '... God is going to let you go to *His* school—where He teaches all sorts of beautiful things to people ... It is called the School of Pain ... And the place where the lessons are to be learned is this room of yours' (Susan Coolidge (Sarah Chauncey Woolsey, 1835–1905), *What Katy Did* (1872), ch. 9, in which a crippled girl is comforting a younger cousin recently—though temporarily—crippled herself). This American story became popular in Britain in the 1890s, when ACD probably obtained it for his daughter Mary. The choice of imagery by Holmes refers to Grace Dunbar's being a governess.

THE CREEPING MAN

First published in the *Strand Magazine*, 65 (Mar. 1923), 211–24, with 5 illustrations by Howard K. Elcock. First American publication in *Hearst's International Magazine*, New York, 43 (Mar. 1923), 8–13, 116, 118, 120, with 6 illustrations by Frederick Dorr Steele.

51 *When I arrived at Baker Street*: although Watson is a visitor to Baker Street in 'The Mazarin Stone', this is the first formal statement that he has left it, and this time for good. He is living in Queen Anne Street in 'The Illustrious Client', while 'The Three Gables' speaks of their former living-room as 'his [Holmes's] room'. Watson is also a visitor in 'The Veiled Lodger'. He resides in Baker Street for the other five Watson-narrated stories in this collection. We learn nothing of the reasons for his departure, except for the mention of a wife (who cannot be Mary Morstan, whom he married at the end of *The Sign of the Four*) in 'The Blanched Soldier'; and while

we know that he is still in medical practice, no information is given about patients, neighbourhood, etc. as in the *Adventures* and the *Memoirs*. The ambience of Watson's home life, so conducive to identification for the reader of the early short stories, is thus reversed: what was once our ground for greatest assurance is now the point where we are most at a loss. The effect is further to disorient Holmes's world in relation to himself in the *Case-Book*. Watson's reticence and secretiveness dissolve the former compact with the reader and help to isolate *us* as well as Holmes. The familiar—in all senses—has abandoned us in pursuit of the unknown.

monograph: a treatise or paper written on one particular subject, normally of interest only to specialists.

the Copper Beeches: In the story of 'The Copper Beeches' (*Adventures*) the spoilt and odious son of Jephro Rucastle suggested a sadistic example from his father.

52 *a tangled skein*: ACD's original title for *A Study in Scarlet*.

wolf-hound, Roy: a dog of large size, formerly used for hunting wolves. Roy was the name of ACD's dog in 1905.

Camford physiologist: the name of the fictitious university is made up from Cambridge and Oxford; a physiologist is a scientist who studies the life processes of living bodies.

53 *the chair of Comparative Anatomy*: there was no such chair at Cambridge at this time, but there was one at Oxford. ACD's knowledge of it as an academic subject derived from Edinburgh in the 1870s.

54 *the E.C. mark*: E.C. stands for East Central (District of London).

55 *cannula*: a tube to be inserted into a cavity.

56 *pathologist*: a scientific student of disease and abnormality; also a person qualified to conduct *post-mortem* examinations.

Lumbago: a condition in which the muscles of the loins are subject to rheumatic pains.

57 *alienist*: a specialist in the diagnosis and treatment of mental illness.

share certificates: documents certifying the holder's ownership of shares in a company or business venture.

57 *Jack*: Bennett has previously been 'Trevor', but that could be a second surname, of ultimately maternal ancestry (cf. 'Conan').

60 *August 25th*: '26th' in other printings, but internal computations turning on 9-day intervals require the amendment.

wistaria: as in the original draft of what became 'Wisteria Lodge' (*His Last Bow*), confirming that the revised spelling in that case was a deliberate contrivance to suggest 'mystery', 'hysteria', etc.

61 *his senseless rage*: although this is presented as proof of Presbury's abnormality, it is strongly reminiscent of Professor Challenger's first meeting with Edward Dunn Malone in *The Lost World*. Challenger, in appearance and in 'some of his peculiarities', was modelled on William Rutherford (1839–99), Professor of the Institute of Medicine at Edinburgh University from 1874 (see ACD, *Memories and Adventures* (1930), 32), though Challenger has nothing of Rutherford's sexual obsessiveness. He ultimately became insane in the late 1880s after denouncing and driving from the university a male student whom he accused of making obscene gestures at him in class (the youth seems to have been quite innocent). Despite attempts to oust him, he retained his chair while in a mental home and eventually returned to it, dying in harness. His predecessor was John Hughes Bennett (1812–75), who bore the same first and last names as Presbury's assistant (with a Welsh middle name in both cases). He is cited in ACD's Edinburgh doctoral dissertation (1885). ACD would have been preparing his reminiscences of medical school while writing 'The Creeping Man': it appears that the story embodies things he knew but chose not to recollect publicly. Although the text speaks of 'a row of ancient colleges', the story shows no real sense of college life at Oxford or Cambridge; instead, it divides Presbury's activities between home and lectures, as it would have been in a university like Edinburgh.

63 *Slavonic*: of a branch of the Slav speech of Eastern Europe.

64 *Commercial Road*: a road in London's East End, off the Docks.

Bohemian: in this case, literally from Bohemia in Central Europe, where Prague is situated.

Mercer: perhaps part of Holmes's 'small, but very efficient organization' ('Lady Frances Carfax', *His Last Bow*).

famous vintage: port wine.

65 *this charming town*: Cambridge, in 'The Missing Three-Quarter' (*Return*), is 'this inhospitable town'.

68 *ramped*: reared up.

Newfoundland: a large bushy-coated dog, often used for hunting. Nana, the Darlings' nurse in J. M. Barrie's play *Peter Pan* (1904), is a Newfoundland dog.

69 *phial*: a small bottle or tube with stopper, for storing chemicals or medicines.

Austrian stamp: Prague was at this presumed time within the borders of the Austro-Hungarian empire.

serum: a watery fluid, produced when blood coagulates and separates.

Anthropoid: a tail-less, man-like monkey or ape.

black-faced Langur: a medium-sized climbing monkey from India.

70 *the elixir of life*: a potion with the supposed power of prolonging life indefinitely.

may not our poor world become: ACD informed several interviewers on his US tour in April and May 1923 that he would write no more Holmes stories, his commitments to Spiritualism being too great to give up time for fictional creation. Perhaps this elegiac note at the conclusion of 'The Creeping Man' was intended in part as a valediction. 'The Sussex Vampire' must therefore have virtually forced itself on ACD, the idea being just too good for him to ignore.

71 *his smell*: expert opinion is that treatment with monkey-serum would not make a human being behave or smell like a monkey. The story is unusually metaphorical.

THE SUSSEX VAMPIRE

First published in the *Strand Magazine*, 67 (Jan. 1924), 3–13, with 4 illustrations by Howard K. Elcock. First American publication in *Hearst's International Magazine*, New York, 45 (Jan. 1924), 30–6, with 4 illustrations by W. T. Benda.

72 *Old Jewry*: in the City of London, on the north side of Cheapside, where the original synagogue of the Jews was erected. They were expelled from there in 1291.

72 *Mincing Lane*: named after houses which belonged to the Minchers, or nuns of St Helens.

per E.J.C.: i.e. written by a member of the firm, probably a partner concealing his identity by this conventional, if irritating, practice.

the giant rat of Sumatra: possibly *Rhizonys sumatrensis*, the great Sumatran bamboo rat. Its connection with 'the assessment of machinery' is obscure.

a Grimm's fairy tale: the brothers Jacob (1785–1863) and Wilhelm (1786–1859) Grimm, in the early nineteenth century, compiled and published a collection of fairy-tales.

73 *'Voyage of the Gloria Scott'*: the story of the *Gloria Scott* is told in the *Memoirs*. It consists almost entirely of narrative by Holmes himself.

Victor Lynch: notice Holmes's curious system of indexing.

Yeggman: safe-breaker (US underworld slang).

Vigor: unknown to other encyclopaedias; possibly a professional strongman.

Vampirism: cf. the popular novel *Dracula* (1897) by Bram Stoker (1847–1912) and also the female vampire in *Carmilla* (1871) by J. Sheridan Le Fanu (1814–73).

Transylvania: a partly mountainous area in the north of Romania (formerly in Hungary).

pure lunacy: Friedrich Murnau (1889–1931) directed the sensational *Nosferatu* (1922), a silent film deriving without acknowledgement from Stoker's *Dracula* and starring Max Schreck (1879–1936). Holmes's comment may perhaps reflect ACD's reaction to it.

This agency: referred to twice in this story, but nowhere else. Its capitalization in English editions is erroneous. American editors recognized it as a generic description, as opposed to Holmes's turning himself into an English Pinkerton's.

Horsham: north Sussex town.

74 *alien religion*: although ACD had ceased to be a Roman Catholic he was as irritated as any former coreligionist at the predelictions which many English Protestants ascribed to Roman Catholics. G. K. Chesterton (1874–1936), in 'The God

of the Gongs' (*The Wisdom of Father Brown*, 1914), describes Italians as being credited with making human sacrifices because of their Catholicism; and while ACD labours the point less, there is a clear implication that Ferguson imagines Catholics have a greater potential for vampirism than Protestants—possibly a vague extrapolation from the doctrine of transubstantiation, which holds that the wine during Mass turns into Christ's blood at the consecration. The same form of prejudice credited Jews with the murder of Christian children for ritual sacrifice: 'alien' at this time was frequently used as a synonym for 'Jew'.

76 *Blackheath . . . three-quarter for Richmond*: Blackheath, for which Watson played, was an amateur Rugby club which in 1871 was one of the founders of the Rugby Football Union. Richmond was another club in the same area, Thames-side London suburbia, one east of London, one west.

a wire: a telegram.

the Old Deer Park: at Richmond.

80 *towering Tudor chimneys*: a feature of the design of houses built during the period 1485–1603.

lichen-spotted: lichen is a type of fungus or moss.

rebus: the carved, moulded, or drawn representation of a name by pictures which punningly depict its component parts. Popular in Edwardian magazines.

half-panelled: panelled in wood from the floor, but only as high as the frieze.

yeoman farmer: a farmer who owns his own land, much idealized in politics and popular literature.

its tail was on the ground: spaniels of all types usually have their tails shortened to 2 or 3 inches, so that the position this spaniel assumed was odd.

84 *a weak spine*: yet we are told that he 'rushed forward'. A spinal injury would probably involve complete paralysis below the waist.

85 *his furrowed forehead*: 'and no wonder,' commented Michael Hardwick on the client's reaction to the previous paragraph (*Complete Guide to Sherlock Holmes*, 191). It is surely objective

rather than subjective to see in such passages the inspiration for:

> 'Precisely, sir. You imply that Miss Pyke's criticisms will have been instrumental in moving the hitherto unformulated dissatisfaction from the subconscious to the conscious mind.'
>
> 'Once again, Jeeves?' I said, trying to grab it as it came off the bat, but missing it by several yards.
>
> (P. G. Wodehouse, 'Jeeves and the Old School-Chum', *Strand*, Feb. 1930)

86 *a Queen in English History*: Holmes refers to Eleanor of Castile (?1245–90), Queen of England (1272–90), who is said to have sucked poison in 1272 from the wounded arm of her husband, Edward I; but this seems to be an unsubstantiated legend and, significantly in this context, Spanish in origin. A possible origin of the incident in the story (suggested by Norman Rosenbaum, Barbara Roden, and Richard Lancelyn Green) is Joseph Bell, who in 1864 saved the lives of several children during a diptheria epidemic by sucking out 'the diseased poisonous mass' of the thick grey membrane at the back of their throats. Bell contracted the disease himself as a result of his heroic conduct. When Queen Victoria was told of the incident sixteen years later by Dr P. H. Watson, she promptly insisted on his ward being renamed after herself on 25 Aug. 1881, while ACD was still in Edinburgh.

bird-bow: a small bow mainly used for killing birds.

87 *curare*: a paralysing South American poison extracted from the woorali root or from the bark of the *Strychnos toxifera*. See Tennyson's poem 'In the Children's Hospital' (1880).

88 *a year at sea*: presumably to 'make a man' of Jacky. The story may have been partly inspired by the case of Constance Kent, believed to have abstracted her baby step-brother from his cradle and cut his throat, in 1860. She was discharged and became a nun, but later she confessed to the murder and was imprisoned 1865–85, after which she may have emigrated to San Francisco. She was about 16 when her step-brother was murdered. Inspector Jonathan Whicher (d. 1871) charged her with the crime in 1860 but could not get a committal for trial. Whicher was a model for Sergeant Richard Cuff in Wilkie Collins's *The Moonstone* (1868).

THE THREE GARRIDEBS

First published in the *Strand Magazine*, 69 (Jan. 1925), 3–14, with 5 illustrations by Howard K. Elcock. First American publication in *Collier's Weekly Magazine*, New York, 74 (25 Oct. 1924), 5–7, 36–7, with 3 illustrations by John Richard Flanagan.

89 *the same month that Holmes refused a knighthood*: the date being 'the latter end of June 1902', this offer can be supposed to have been part of Lord Salisbury's resignation honours or Edward VII's coronation honours, as ACD's was.

telephone directory: telephones do not appear at 221B Baker Street until the *Case-Book* stories, although the telephone was introduced into London in 1876 and Athelney Jones makes use of police telephone services in *The Sign of the Four*. A telephone number first appeared on ACD's personal writing-paper in 1908, after the move to Crowborough, but he took some time to think of it automatically since in 'The Bruce-Partington Plans' (*His Last Bow*) Holmes must still send messages from the field to Watson in Baker Street by telegram.

90 *Counsellor at Law*: the American equivalent of an English barrister.

Moorville, Kansas, USA: no such place, apparently, in Kansas; but there is a Moorsville in Alabama and another in Indiana.

any eccentricity of speech: but he *does* use several Americanisms (or quasi-Americanisms), noted below.

91 *fool trick*: (American) a stupid breach of existing understanding.

that puts it different: (American) throws a different light on the matter.

92 *Alexander Hamilton Garrideb*: Alexander Hamilton (1755–1804), the USA's first Secretary of the Treasury, who laid the foundation of its national economic policy in 1789–95, was a natural choice of name for Federalist-Whig parents in the early nineteenth century (e.g. Alexander Hamilton Stephens (1812–83), Vice-President of the Confederacy). As an invention, it would have borne some smack of authenticity to an educated Englishman of the day.

wheat pit: (American) a market where wheat and other grain are traded; hence the muckraking novel by Frank Norris

(1870–1902), *The Pit* (1903), which ACD thought to be 'one of the finest American novels' (*Through the Magic Door*, 44).

92 *Arkansas River*: major river flowing from Colorado to Arkansas, via Kansas.

Fort Dodge: a conflation of Fort Dodge, Iowa, over 300 miles from the Arkansas River, and Dodge City, Kansas, which meets the description. ACD would have worked this out with a map in front of him, as he used to do with London before he settled there, so the error would seem deliberate and a further clue to John 'Garrideb's' not being the Kansan he claims to be. A Chicagoan would know Iowa far better than Kansas, and might easily drop the Iowa name in place of a similar one.

lumber-land: land used for forestry and the growing of trees for timber.

Topeka: capital of Kansas in the east of the state.

pan out: (American) work out, presumably originating in the sifting of gold deposits by panning, when prospecting.

93 *vacancy*: the word employed in 'The Red-Headed League' (*Adventures*) in the comparable plot to get Jabez Wilson out of his premises.

agony columns: personal messages or advertisements in newspapers, usually intimate.

old Dr Lysander Starr: there is a Colonel Lysander Stark in 'The Engineer's Thumb' (*Adventures*) and, in real life, Dr Leander Starr Jameson (1853–1917), of Jameson Raid fame, who was strongly condemned by ACD in *The Great Boer War* (1900).

94 *covert for putting up a bird*: a covert is a bush or thicket used as cover by game birds. They have to be disturbed ('put up') so that the hunter may get a shot at them.

syncopated: one-sided, and hence abridged.

a lovely spring evening: but at the beginning of the story we are told that the adventure took place in late June.

Tyburn Tree: the gallows on which public executions took place until 1759, situated at the meeting of Edgware Road and Oxford Street (the site of the present Marble Arch).

Early Georgian: the first four Georges reigned from 1714 to 1830.

95 *cadaverous*: lean, gaunt, haggard; faintly corpse-like.

flint instruments: weapons and tools of the kind used by primitive man.

'Neanderthal', 'Heidelberg', 'Cromagnon': believed to be different stages of human evolution.

Syracusan: Syracuse was a Greek colony on the eastern coast of Sicily which flourished between the eighth and third centuries BC.

96 *Alexandrian school*: ancient Alexandria was a great city and seaport on the Nile delta in Egypt, founded in 332 BC by Alexander the Great.

Sotheby's or Christie's: two of London's great auction houses.

Hans Sloane: Sir Hans Sloane (1660–1753), naturalist, physician, collector, and benefactor; responsible for the founding of the British Museum.

97 *our telephone appointment*: 'It is an odd point that when Holmes was ringing up Mr Nathan Garrideb, he advised him not to mention to the "American lawyer" that they were coming; yet when they arrived he asked "Did you tell him of our telephone appointment?" and appeared in no way surprised by an affirmative answer. Had he already summed up the eccentric collector as a hopeless nincompoop who could be assumed to be incapable of carrying out instructions?' (Dakin, 270) This would account for the decision not to tell Nathan Garrideb the truth and allow him to go in quest of Howard Garrideb in Birmingham, for which reticence some justification seems necessary as the quest drove Nathan insane.

98 *drills*: machines for sowing seeds in rows.

harrows: implements for breaking up and levelling land ready for planting with new seed.

buckboards: four-wheeled carriages of a type common in the USA during the nineteenth century; particularly popular in the country.

Artesian Wells: boreholes to tap water, named after the region of Artois in northern France.

Aston: a large suburb of Birmingham (England).

Birmingham: also used as a place to which to send a dupe in 'The Stockbroker's Clerk' (*Memoirs*). ACD was apprenticed to a doctor there in his student days.

98 *figured out*: (American) worked out.

connections: times of trains, routes, stations.

100 *Queen Anne*: reigned from 1702 to 1714.

101 *Newgate Calendar*: a publication started in 1773 which catalogued the famous inmates of London's Newgate prison and gave lurid details of their crimes.

friend Lestrade: Lestrade does not play a leading part in the *Case-Book*. Why did Holmes not enlist his help with the arrest of Killer Evans, after the help Lestrade had already given?

Yard: Scotland Yard, the name by which London's Metropolitan Police Force (especially the Criminal Investigation Department) was popularly known; derived from the location of the building in which the Force was housed up to 1890.

Rogues' Portrait Gallery: popular name for a collection of pictures of criminals kept at Scotland Yard.

Rodger Prescott: 'Presbury' in the *Strand*; obviously corrected when the duplication of the name with that of the professor in 'The Creeping Man' was realized. ACD always spelt 'Rodger' thus, the spelling being common as a last name in Edinburgh (where 'Roger' as a last name is rare).

102 *devilish ingenuity*: it is hard to credit Evans with such ingenuity. Gavin Brend, in *My Dear Holmes* (1951), suggests that Evans got the Garrideb idea from John Clay of 'The Red-Headed League' (*Adventures*).

103 *jemmy*: a short crowbar used by burglars.

played me for a sucker: (American) fooled me, made me the victim of your deception (often used when the victim sought to do the deceiving).

105 *crazy boob*: (American) implying imbecility (like 'old stiff' at the end of the paragraph).

Where do you get me?: (American) 'What offence do you charge me with?'

Brixton: a district of London south of the Thames.

Bench: the official seat of a judge (here used to mean the judge himself).

shades . . . just emerged: probably inspired by the popular story of Thomas Carlyle's comment on the dismissal of his unsatisfactory maid-servants: 'The demons return to the shades from which they should never have emerged.'

THE ILLUSTRIOUS CLIENT

First published in the *Strand Magazine*, 69 (Feb./Mar. 1925), 100–18, 259–66, with 4 illustrations by Howard K. Elcock. First American publication in *Collier's Weekly Magazine*, New York, 74 (8 Nov. 1924), 5–7, 30, 32, 34, with 4 illustrations by John Richard Flanagan.

A number of the following notes relating to Chinese ceramics have been based on comments from Professor Bonnie S. McDougall, Department of East Asian Studies, University of Edinburgh and in all cases are followed by the initials 'BSM'; matter directly supplied by Professor McDougall has been placed in quotation marks. The editor and general editor are indebted to Professor McDougall for her assistance in establishing the authenticity of these allusions.

106 *the supreme moment of my friend's career*: this has elicited adverse comments—justifiably, if it means the issues or the client, since Holmes had dealt with greater things in each category. But it is difficult to think of a confrontation (other than at the Reichenbach Falls) possessed of greater drama than the present story's climax.

the Turkish Bath: a type of hot-air bath of varieties of treatment; it was first popular in the Middle East.

Northumberland Avenue: a fashionable street in Westminster, linking the Victoria Embankment and Trafalgar Square.

September 3: Oliver Cromwell's lucky day, on which he won the battles of Dunbar (1650) and Worcester (1651); he also died on this day (1658), as ACD well knew. A crypto-ironic choice of date for the first tidings of the Illustrious Client.

the Carlton Club: the leading Conservative political club.

107 *Sir George Lewis*: 1833–1911, of Lewis and Lewis, the most illustrious, and probably the most discreet, solicitor of his time.

Queen Anne Street . . . Baker Street: both in the heart of London's West End. Holmes's rooms were on an upper floor of 221B Baker Street (a fictitious address); Queen Anne Street is a doctors' quarter.

107 *Colonel Sir James Damery*: Major-General Sir Henry Ponsonby (1825–95), private secretary to Queen Victoria (whence 'Damery', or business of the *dame*), of an Irish aristocratic family (the Earls of Bessborough), seems an obvious historical antecedent (though probably personally unknown to ACD). It would have made sense to use so obviously deceased a model, easily identifiable in court circles, since it is quite possible Damery may also embody a caricature of some actual courtier. It is a vivid portrait, and the closer the scrutiny, the less flattering to any original it becomes. Damery's own flattery may impose on Watson (to the extent of inflating his descriptive rhetoric well beyond the norm), and possibly even on Holmes—who notoriously liked applause; but as a diplomat he is pointedly presented as being almost self-destructive. His mission is (a) to hire Holmes, (b) to conceal the identity of his client. He almost loses an otherwise willing Holmes by his absolute insistence on avoiding the saving devices Holmes offers him—to let himself be taken for the client, or to imply that the client is de Merville. It is clear, in fact, that Damery is desperately anxious to insist that there is a client whom he cannot name, while doing everything to proclaim his identity, including leaving the brougham with armorial bearings standing outside Holmes's door for the edification of the Baron's agents, any passing assassin, or the press. This is evidently quite deliberate on the author's part and may reflect recent problems with intermediaries when ACD was asked to contribute a piece for *The Book of the Queen's Doll's House* (for which he wrote 'How Watson Learned the Trick': see *The Return of Sherlock Holmes* in the present series). The book appeared in June 1924; contributors were originally approached at the request of HRH Prince Marie Louise (1872–1956), no doubt via a courtly emissary. ACD's comment that the story 'moves adequately in lofty circles' (*Strand*, June 1927, quoted by Richard Lancelyn Green, *Uncollected Sherlock Holmes*, 323) shows no decline in his sense of humour.

cravat: a neckcloth.

spats: a kind of gaiter worn round the ankle and above the shoe.

no more dangerous man in Europe: possibly prompted by a story at the close of the official life of Benjamin Disraeli (1804–81) by W. F. Monypenny and G. E. Buckle: 'Disraeli was . . .

always elaborately civil to the Gladstone family. Talking to one of the daughters at some reception, where one of the principal guests was a foreign diplomatist of very varied political career, "That", he said in response to her inquiry, "is the most dangerous statesman in Europe—except, as your father would say, myself, or, as I should prefer to put it, your father." ' There is some similarity in the relative situations of Gladstone, Disraeli, and Queen Victoria with those of Holmes, Gruner, and Violet de Merville, and Gruner has several touches of Disraeli.

the late Professor Moriarty: Holmes's most famous antagonist. See 'The Final Problem' (*Memoirs*), 'The Empty House' (*Return*), and *The Valley of Fear*.

the living Colonel Sebastian Moran: see 'The Empty House'. Moran presumably got penal servitude for the murder of Ronald Adair.

108 *kid-gloved hands*: it is curious that Sir James did not remove his gloves. He certainly expects his agents to do so.

Prague: then in the Austrian Empire, now the present capital of the Czech Republic. Why should a Prague court have had jurisdiction over a crime committed on the Swiss–Trentino (then Austrian) frontier? Possibly an arrest in Prague is presumed, followed by the Baron's public examination and discharge for lack of evidence.

the Splügen Pass: a high Alpine pass between Italy and Switzerland.

109 *Khyber*: a mountain pass providing access from Pakistan (still India in Holmes's day) to Afghanistan. Military campaigns involving the British took place in 1839–42, 1879–80 (the Second Afghan War), and 1897. Watson met Holmes while recovering from a wound received in the Second Afghan War.

110 *Kingston*: Kingston-on-Thames, a royal borough, situated in the south-west suburbs of London.

111 *Hurlingham*: where there were grounds used by the Polo Club; situated on the banks of the Thames, near Fulham in West London.

Charlie Peace: a famous nineteenth-century criminal, born in 1832 and executed for murder in 1879. He appeared on stage as 'the modern Paganini' playing a one-stringed violin.

111 *Wainwright*: Thomas Griffiths Wainwright (1794–1837), artist and poisoner. See Oscar Wilde's 'Pen, Pencil and Poison: A Study in Green' (1889), reprinted in his *Intentions* (1891).

Shinwell Johnson: perhaps part of that 'small, but very efficient organization' to which Watson refers in 'Lady Frances Carfax' (*His Last Bow*).

Parkhurst: a prison on the Isle of Wight, off the south coast of England.

112 '*nark*': slang for police spy or informer.

Simpson's: a famous restaurant in the Strand.

the Strand: the old road linking the City of London with the West End.

114 *Apaches*: street thugs, ruffians.

Montmartre: artistic district in the north of Paris.

115 *scorbutic*: with a face disfigured by blotches or scabs.

brand: torch.

leprous: leprosy was the name given to a whole variety of skin diseases at this time.

yours to the rattle: yours till death. The 'death-rattle' can be heard in the throat.

116 *It's a book he has*: perhaps inspired by the 'Black Diary' attributed to Sir Roger Casement (1864–1916) and circulated to persons such as ACD who supported calls for Casement's reprieve after he was convicted for high treason. ACD contemptuously pointed out the irrelevance of such evidence of supposed sexual (in this case homosexual) obsession to the graver charge at issue, but the success of the 'Black Diary' campaign in discouraging others who initially favoured clemency towards Casement may have given him the idea of the effectiveness of such a weapon. Gruner, however, has no resemblance to Casement, on whom ACD drew for the character of Lord John Roxton in *The Lost World* (1912).

121 *Grand Hotel*: formerly in Trafalgar Square.

Charing Cross Station: built in 1864; one of the two London termini of the South Eastern Railway.

I think: the first of the two instalments of the story in the *Strand* finished with the placard.

the Café Royal: opened in 1865, it became extremely popular with artists and writers.

Charing Cross Hospital: situated on the north side of the Strand, near Charing Cross Station. Dr Mortimer (*Hound*) was once on its staff.

122 *hansom*: a two-wheeled carriage drawn by a single horse; named after J. A. Hansom (1803–82).

brougham: a four-wheeled carriage, named after Henry, Lord Brougham (1778–1868), Lord Chancellor (1830–4).

Morphine: a drug derived from opium, used as a painkiller.

compress: a surgical pad used to apply pressure.

single-stick: a heavy wooden stick, some three feet long, with a protected handle at one end.

123 *tobacco-slipper*: Holmes kept his tobacco in the toe-end of a Persian slipper.

erysipelas: an inflammatory skin disease, usually facial.

the Cunard boat: the Cunard line was founded in 1838 by Samuel Cunard.

Ruritania: an imaginary liner, named after the fictitious Central European kingdom in Anthony Hope's novel *The Prisoner of Zenda* (1894).

124 *the London Library*: a major library from which books can be borrowed by subscribers from an extensive range of general and specialist works.

St James's Square: dating from the time of Charles II.

a goodly volume: unidentified; perhaps it contained information not in standard reference works (BSM).

hall-marks of the great artist-decorators: if potters, these would probably not have had hallmarks. It was not the custom of individual potters to sign their names, but it was common for the kiln or manufacturer's name to be given along with the name of the dynasty and the reign title (BSM).

Hung-wu . . . Yuan: the references to Hung-wu (reign name), Yung-lo (reign name), Sung (dynasty), and Yuan (dynasty) are correctly given and relevant to the subject. 'Tang-ying' is better written as 'T'ang Ying' (BSM).

125 *egg-shell pottery*: originated during the Ming (1368–1644), though it was perfected in, and is more commonly associated with, the Ch'ing (1644–1912). But (1) it is more likely to be called porcelain than pottery; (2) if monochrome, it would be unlikely to be a heavily saturated colour like 'deep blue'; (3) 'delicate little saucers' were not an indigenous form of pottery. A set of six could have been made for export (what happened to the cups?), but they would not have been found in the imperial palace of Peking. Perhaps ACD had in mind a kind of small bowl (BSM).

Dr Hill Barton: the name derives from the historian John Hill Burton (1809–81), with whose family the boy ACD was intimate in Edinburgh.

Christie or Sotheby: see note to 'The Three Garridebs', p. 255.

South African gold king: a man who had made his fortune in the South African gold-mining industry.

the great boom: the sudden rush of businessmen and miners to the Transvaal province of South Africa when large deposits of gold were discovered there in 1885.

a plush-clad footman: plush is a soft velvety cotton fabric. A footman was a servant to lend ostentation to the reception of guests and to waiting at table.

a small brown vase: this could have been T'ang and from the seventh century, but the description 'richer glaze' does not sound convincing about T'ang porcelain (or pottery) (BSM).

128 *Shomu . . . Shoso-in . . . Nara*: 'The baron's questions are well chosen: an impostor who had read only one book on Chinese pottery would very probably have not been able to answer them.' Nara was an early capital of Japan in the seventh and eighth centuries AD. The late Nara period, and especially the reign of the emperor Shomu (724–49), is known for its flourishing art and architecture. The Shoso-in is a storehouse for Buddhist ritual objects and the personal belongings of Shomu (BSM).

the Northern Wei dynasty: founded in 386 by a Central Asian people known as the T'o-pa. They conquered all of North China, but in 535 their state was split into two halves that had both disintegrated by 557. The Northern Wei dynasty is noted for its Buddhist sculpture, especially in the 'caves' at Yunkang and Lungmen outside its early and late capitals. The ceramics of the Northern Wei appear to have been unremarkable. A kind of proto-porcelain existed in the period of division, but true procelain had not yet been developed. Again, a fairly searching question, and it is not surprising that Watson is unable to answer it (BSM).

129 *vitriol*: concentrated sulphuric acid.

130 *a hypodermic of morphia*: why did Watson have this on his person (and oil and wadding) when he expected only a discussion on ceramics? Presumably, he had brought his medical bag with him to add verisimilitude to 'Dr Hill Barton'; the items themselves would have been useful when visiting the convalescent and hyperactive Holmes.

The wages of sin: 'The wages of sin is death' (Romans 6: 23).

132 *As I was myself overdue*: was this at the bedside of a patient (hence the bag)? Dr Verner had bought Watson's practice, but Queen Anne Street suggests a new one.

cockaded coachman: the coachman's uniform hat distinguished by a rosette or plume.

armorial bearings: the coat of arms displayed on the outside of the brougham's door panel.

THE THREE GABLES

First published in the *Strand Magazine*, 72 (Oct. 1920), 319–28, with 4 illustrations by Howard K. Elcock. First American publication in *Liberty*, New York, 3 (18 Sept. 1926), 9–14, with 6 illustrations by Frederick Dorr Steele.

133 *Three Gables*: a gable is the triangular upper part of the exterior wall of a building, at the end of a ridged roof.

negro: uncapitalized in all save most recent editions, as normal at this time. The use of the capital N was a statement of racial egalitarianism in the USA up to 1965.

Masser Holmes . . . Masser Holmes . . . Masser Holmes: the second and third 'Masser' became 'Mr' in the *Strand*, inconsistently with other occurrences in the story—whether as a result of carelessness or sensitivity cannot now be ascertained. 'Masser' (Master or Mister) was the standard Negro dialect used (mostly by whites) on the music-hall stage or in stories about slavery in the American South.

134 *the Holborn Bar*: probably fictitious.

the Bull Ring in Birmingham: a large open space in the centre of Birmingham.

135 *Spencer John*: if Spencer John is the gang-leader, he must have been as remote from real power as the later Merovingian French kings. He is apparently told nothing of the work of Barney and Susan Stockdale for Isadora Klein, and in any case it is difficult to see the possessor of so aristocratic a name establishing authority over such coarse characters. Donald A. Redmond (*Sherlock Holmes: A Study in Sources*) identified a 'Sir Spencer John who flourished about 1834, but who . . . cannot be traced further' (220). Did this story start life as a Regency romance, describing the suppression of a *roman à clef* dealing with swell mobster violence on a discarded lover? Certainly such a *roman* would have had far more impact at the beginning of the nineteenth century than at the end (by which time every 'prentice hand claimed truth for fictions, as ACD in his own case bore witness). The story itself, apart from its odious racist language and vulgar manners, is strangely inconsistent with Holmes and Watson in its context. They talk familiarly of underground gangs, and Holmes is cheek-by-jowl with the chief gossip columnist (whose headquarters again have a Regency ring and would be of much more limited use in 1900); he also as a matter of course knows young men-about-town whom 'all London knew'—people towards whom he was usually completely indifferent. As a story, 'The Three Gables' needs very little change to be about figures entirely different from Holmes and Watson. Perhaps it was originally about different characters, and then a Holmes deadline threatened for the first of the series of six.

Harrow Weald: a suburb to the north of Harrow; a weald is a tract of open or wooded country.

136 *Attaché*: a member of the diplomatic service attached to an embassy abroad.

138 *Paregoric*: a solution of opium, benzoic acid, camphor, and oil of amber in alcohol, used as a painkiller. Holmes probably suggested it as a remedy for Susan's wheeziness. Sir Henry Irving's appearance in ACD's play *Waterloo*, about a very old

soldier, was reviewed by Bernard Shaw in the *Saturday Review* (11 May 1895) under the derisive title 'MR IRVING TAKES PAREGORIC'.

139 *a Raphael*: Raffaello Sanzio (1483–1520), Italian Renaissance painter.

a first folio Shakespeare: the First Folio (printed in 1623) is the name by which the earliest collected edition of Shakespeare's plays is known.

a Crown Derby tea-set: a tea-set of fine British porcelain, made in the town of Derby, and so called because of the incorporation of a crown in the potter's identification mark.

lock, stock, and barrel: altogether. The phrase related to the parts of an old-fashioned rifle.

140 *Auctioneer and Valuer*: an expert who, for a fee, assesses the value and conducts auctions of his client's property.

'Milano.' 'Lucerne': respectively, the principal city of Northern Italy and a popular tourist centre in Switzerland.

141 *Langdale Pike*: the Langdale Pikes are two hills in what was formerly Westmorland, overlooking Wordsworth's Grasmere. D. M. Dakin remarks on the improbability of the name; but perhaps it was his *nom de plume* (like the columnist 'Peterborough' in the *Daily Telegraph*).

142 *a broken reed*: 'Thou trustest in the staff of this broken reed' (Isaiah 36: 6).

143 *chloroformed*: chloroform ($CHCl_3$) is a colourless liquid with a characteristic odour. It was first developed in Edinburgh Medical School from 1847 to ease the pains of childbirth.

145 *mighty poor stuff*: the Inspector was surely right.

146 *Grosvenor Square*: in the West End of London, to the south of Oxford Street.

conquistadores: the Spanish conquerors of Central and South America.

Pernambuco: a seaport on the easternmost edge of Brazil, now known as Recife.

'*belle dame sans merci*': best known from Keats's poem of that name, beginning 'O what can ail thee, knight-at-arms'; originally from the medieval poem by Alain Chartier.

Duke of Lomond: a fictitious title. Loch Lomond is a famous lake in Scotland.

147 *Arabian nights*: *The Arabian Nights*, also known as *The Thousand and One Nights*; a celebrated collection of oriental tales.

over-rated: 'underrated' in *Strand* and book editions.

148 *calcined*: charred, consumed by fire.

150 *compound a felony*: see note to 'The Mazarin Stone', p. 14.

THE BLANCHED SOLDIER

First published in the *Strand Magazine*, 72 (Nov. 1926), 423–34, with 5 illustrations by Howard K. Elcock. First American publication in *Liberty*, New York, 3 (16 Oct. 1926), 12–14, 17, 19, 21, with 5 illustrations by Frederick Dorr Steele.

151 The text of the story's first page in the *Strand* surrounded the encircled 'This is the first Adventure ever related by Sherlock Holmes himself', a statement which ignored 'The *Gloria Scott*' and 'The Musgrave Ritual' (*Memoirs*)—both being previous 'Adventures' related by Holmes himself, albeit within a Watsonian framework.

the Boer War: the South African War of 1899–1902, fought between the British and the Boers (South Africans of Dutch descent). ACD himself served with the British forces during the war and wrote its history (*The Great Boer War*, 1900).

deserted me for a wife: presumably a second wife since Mary [Morstan] Watson had died before 'The Empty House' (fictional date 1894).

152 *Imperial Yeomanry*: a mounted troop called the Yeoman Cavalry volunteered for service in South Africa. During the war, in 1901, its name was changed to the Imperial Yeomanry.

Throgmorton Street: associated with London stockbroking.

martinet: a strict disciplinarian.

153 *the Crimean VC*: the Victoria Cross, instituted by Queen Victoria for acts of bravery during the Crimean War (1854–6). As to Emsworth, the name (but not its possessors) was probably inspired by P. G. Wodehouse's Lord Emsworth, launched in *Something Fresh* (1915).

Diamond Hill . . . Pretoria: a ridge twenty miles east of the capital of the Boer Republic of the Transvaal.

Cape Town: chief town of the British territories in South Africa at the time of the Boer War.

Southampton: the sea-port midway on the south coast of England that was the chief disembarkation point for troops returning from South Africa.

154 *Bedford*: market town in south central England.

trap: a light, open, two-wheeled carriage.

half-timbered Elizabethan foundation: the oldest part of Tuxbury Old Hall went back to the reign of Elizabeth I (1558–1603). Half-timbered walls were walls in which some of the wooden supports had been left exposed, the spaces between them being filled with plaster or brickwork.

portico: a range of columns above the front or side of a building.

butler: chief manservant in a household, with responsibility for overseeing the other members of the staff, above all the wine-steward. A butler would conventionally be known by his last name: the use of his Christian name indicates an old and close tie and a reclusive family.

a barney: a quarrel.

155 *a straggling grey beard*: cf. Wodehouse's short story 'Lord Emsworth Acts for the Best' (*Strand*, June 1926; *Liberty*, 5 June 1926), which features Lord Emsworth bearded—uniquely so in the Emsworth saga. Freddie Threepwood, Lord Emsworth's son, threatens 'to retire to some quiet spot [possibly the ancestral home, Blandings Castle] and there pass the few remaining years of my existence, a blighted wreck'. A more startling instance of the affinity of Wodehouse's work to 'The Blanched Soldier' is noted below; as in that instance, the consonance seems impossible to explain, given the time-scale, except by a previous exchange of ideas between ACD, the old master, and his disciple Wodehouse.

156 *veldt*: the open grasslands of South Africa.

158 *standing before me*: cf. Wilkie Collins's *The Woman in White* (1859), end of Part the Second, where the narrator, looking at the

tombstone 'sacred to the memory' of his beloved, looks up to see her 'looking at me over the grave'.

158 *I was some little time before I could throw it up*: 'it was some little time before I could open it' (*Strand*). The revision in all book-texts makes the exit of Dodd much more vivid and is obviously an author's change, revealing that ACD continued to give the Holmes stories scrutiny after their first publication.

159 *a gamekeeper's residence*: a gamekeeper was employed on an estate to rear game birds and protect them from poachers. The most famous gamekeeper in English literature is Mellors in D. H. Lawrence's *Lady Chatterley's Lover* (1928).

161 *the astute reader*: virtually an admission of Watson's irreplaceable value in linking Holmes to his audience—as is the strength of the story's being in the speech of the other characters.

162 *the Abbey School . . . the Duke of Greyminster*: although differing in date, probably the same as the Priory School and the Duke of Holdernesse (see 'The Priory School', *Return*). Possibly a second phase of that case is implied, as Watson was firmly in Baker Street during the present narrative.

the Sultan of Turkey: in 1903 this was Abdul Hamid II ('Abdul the Damned' (1842–1918)), whose other depravities may be very slightly offset by his enjoyment of the Holmes stories.

163 *Euston*: Christopher Morley points out that one cannot get to Bedfordshire via Euston.

166 *Eastern railway line*: ACD, in *The Great Boer War*, describes the fight at Bronkhorst Spruit (Buffelsspruit); his map shows the Delagoa Bay Railway line, which ran east of the city to the ocean.

stoep: (Dutch) steps, i.e. veranda.

169 *Lord Roberts*: Field-Marshal Frederick Roberts, first Earl Roberts (1832–1914), the British commanding general in South Africa in the early phases of the Boer War.

systematized common sense: a bonus of the story is that it gives us this three-word summation of Holmes's method in place of the *magnum opus* 'The Whole Art of Detection' he had intended. His concern for the tricks of his trade in self-revelation is also an intended disenchantment, analogous to an idolized actor proving on personal acquaintance more concerned about his craft than his art.

170 *The Lancet . . . British Medical Journal*: both these medical publications are still in circulation. ACD had written for both in the early 1880s.

171 *dermatologist*: someone who specializes in the treatment of skin diseases.

a physical effect . . . that which it fears: an authoritative medical judgement from ACD's professional experience; but cf. (a) malingering as a monograph topic ('The Dying Detective' [*His Last Bow*]), and (b) 'Men die of the diseases which they have studied most' and that theme's development in 'The Surgeon Talks' (*Round the Red Lamp*, 1894).

this joyous shock: P. G. Wodehouse published 'A Slice of Life' in the *Strand* (Aug. 1926) and *Liberty* (7 Aug. 1926), two months before both magazines carried 'The Blanched Soldier'. The story deals with an imprisoned girl in a country house. Her lover, discovering her in defiance of her formidable guardian (Sir Jasper ffinch-ffarrowmere, Bart.), finds her a prisoner of her own volition after her skin has turned piebald. He cures her with a patent medicine (having caused her condition by the use of another) and also reduces her uncle's obesity and the butler's lumbago (see 'The Creeping Man'). The Wodehouse story could not have inspired ACD, simply for want of time; but they seem to be closely linked and one can only conclude that they resulted from a mutually inspirational discussion. It is possible that 'The Blanched Soldier' may have been in hand for some years. ACD had been promising Greenhough Smith a six-part series as far back as 1921–2 when working on 'Thor Bridge' (MS, Metropolitan Library, Toronto). Perhaps he wrote 'The Blanched Soldier', held back from publishing it, but at some point showed it to Wodehouse. On content, it seems more likely that 'A Slice of Life' (later reprinted in *Meet Mr Mulliner*, 1927) derives from 'The Blanched Soldier' than vice versa.

THE LION'S MANE

First published in the *Strand Magazine*, 72 (Dec. 1926), 539–50, with 3 illustrations by Howard K. Elcock. First American publication in *Liberty*, 3 (27 Nov. 1926), 18, 21, 23, 25–6, 29, 31, with 7 illustrations by Frederick Dorr Steele.

Text: collated with the MS, as published in facsimile, with an introduction by Colin Dexter and an afterword by Richard Lancelyn Green, by Westminster Libraries and the Sherlock Holmes Society of London (1992). Only major divergencies between the MS and printed texts are noted here. Some readings of deleted matter are conjectural.

172 *Ah! had he ... the Lion's Mane*: substituted in MS for this passage:

> It is possible that he would in any case have rejected this case from his records for in his loyalty he would always dwell upon my successes. I do not think that I can look back on the adventure of the Lion's Mane with any particular personal pride and yet in its rarity I place it very high among my collection.

the Downs: the South or Sussex Downs; a stretch of treeless chalk uplands extending along the south coast of England.

coaching establishment: similar to a school, but usually on a smaller scale, where pupils are trained intensively in preparation for public examination or university entrance.

173 *several masters*: corrected in revisal of MS from 'three masters'.

a rowing Blue: at Oxford or Cambridge a Blue is a student who has been selected to represent the university at some sport.

I am a: 'long distance' deleted in MS.

174 *with an eager air of warning*: added in revisal of MS.

which burst in a shriek from his lips: added in revisal of MS.

sense. Then: sentence deleted between these words, in MS: 'Such also was the impression upon my companion.'

air: next words 'gave a terrible shriek' deleted in MS.

Burberry overcoat: a kind of waterproof overcoat.

surds and conic sections: in mathematics a surd is the irrational or inexpressible root of a rational number (for example the square root of 2). A conic section is produced when a cone is cut or divided by a plane.

175 *What can I do?*: substituted in MS for:

> 'Where is the nearest police station?'
> 'There is one at Dove Cove Fulworth.'
> 'Then hurry there, I beg you.'

For which last 'You can hurry to the police station at Fulworth' substituted.

Without a word he made: substituted in MS for 'He hurried'.

the body: The ensuing sentences deleted in MS:

> Dr Mordhouse the well-known naturalist who was summering on the South Coast had joined him. Morning, noon & evening the Doctor's sunhat and butterfly net were familiar objects on the Downs and along the beach. I need not explain that he is probably the most popular living writer upon the subject. What is more important at this crisis is that he was of a cheery sunny disposition a fit comrade for a man in trouble.

No one else had: 'Had' substituted in MS for 'appeared to have'.

by this track: inserted during revisal of MS.

lay: substituted in MS for 'was'.

folded: 'up' deleted in MS as next word.

would seem: substituted in MS for 'seemed'.

entered the water: substituted in MS for 'bathed'.

as I hunted round: inserted in MS on revision.

where the print: changed in MS from 'where his foot print', 'of' from 'first with', 'also of his naked foot' from 'afterwards naked'.

indicated: substituted in MS for 'might indicate'.

done so: Next sentence deleted in MS:

> The reader will see that I take him into my confidence as I go, and have, I fear, none of the literary wiles of Watson.

176 *as strange a . . . confronted me*: interpolated in MS.

naked: interpolated in MS.

suddenly: interpolated in MS.

at any rate without drying himself: interpolated in MS.

was: interpolated in MS.

but . . . concealment: substituted in MS for 'No one could emerge from them now unseen and they would all be examined in turn.'

again: interpolated in MS.

176 *were*: substituted in MS for 'lay'.

roads: ensuing 'to see [?]' deleted in MS; or possibly 'to use'.

obvious: substituted for 'serious' in MS.

was, of course: substituted in MS for 'and Dr Mordhouse were'. Here and at other points 'Mordhurst' appears initially, and then is corrected to 'Mordhouse'. The point is of significance, enabling us to deduce that the Mordhouse version was not a casually discarded draft but had itself been made the subject of careful revision before Mordhouse was obliterated and the story transformed. The original 'Mordhurst' was evidently set aside as too close to 'Stackhurst'. Some of the surviving MS revisions may also have been made while the Mordhouse version was still in use.

drew: substituted in MS for 'took'.

Lewes: municipal borough of Sussex East, on the river Ouse.

also: substituted in MS for 'and'.

fresh: interpolated in MS.

made: substituted in MS for 'left'.

177 *knife*: preceding word 'pocket' deleted in MS revisal.

having first . . . searched: this phrase apparently interpolated during revision of MS.

removed to: ensuing words 'a mortuary' evidently deleted in MS as soon as written.

As I expected: interpolated in MS at some point after interpolation of the ensuing words.

nothing . . . but he: interpolated in MS, substituting for 'He'.

in: substituted in revisal of MS for 'upon', subsequent to the previous correction.

Fulworth: in MS first draft, not a substitution.

the letters: substituted in MS for 'them'.

save, indeed: substituted in MS for 'Save'. In the original version the sentence (as it was at that point) is assigned to Holmes. The revisal of the MS involved various observations being assigned to persons other than their original speakers, with the result that (as critics have complained) Holmes in the final version makes remarks not at all characteristic of him. In the story as initially written, he had not said them.

you: substituted in MS for 'us', the line initially being Stackhurst's.

I remarked: interpolated in MS, and quotation marks then closed here. It is obvious from the story that the swimmers bathe in the nude.

'It is . . . several of the: interpolated in MS in place of continuation of previous speech when Stackhurst's having identical words but with 'my' for 'the'.

algebraic: interpolated in MS.

178 *seem to*: interpolated in MS.

a: substituted for 'his' in MS revision.

draw: substituted in MS for 'attract'.

owns: substituted in MS for 'runs'.

bathing-cots: small boats used by people bathing in the sea.

into Fulworth: 'to Fulworth' interpolated in MS after 'in'.

man did not ill-use: substituted in MS for 'chap did not mistreat'.

scourge: 'wire' deleted in revisal of MS as preceding word.

bay: 'cove' written originally and promptly deleted in favour of 'bay'. Richard Lancelyn Green in the facsimile publication mentions that ACD and his family had spent the summer in Lulworth Cove in Dorset, whence some inspiration would seem to have come.

rising: 'raising' written originally and promptly deleted in favour of 'rising'.

179 *sideways*: the preceding words 'rather malevolent' deleted in revisal of MS.

passed: 'stopped' written originally and promptly deleted in favour of 'passed'.

habitable: 'tolerable' before substitution in MS revisal.

angry eyes: substituted in MS revisal for 'clenched fists'.

forcibly: interpolated in MS.

180 *wide-eyed and intense*: interpolated in MS.

already: interpolated in MS.

Fitzroy: substituted in MS for 'Mr McPherson'.

concentration: substituted in MS for 'intensity'.

It seems: 'A few words' originally written and promptly deleted in favour of sentence beginning thus.

181 *here*: substituted in MS for 'now'.

Fitzroy's: substituted in MS for 'his'.

us from: substituted in MS for 'our'.

182 *anything*: substituted in MS for 'any question'.

was as good as her word: substituted in MS for 'did so'.

, but she admitted: substituted in MS for '. Yes she must admit'.

Fitzroy: substituted in MS for 'Mr McPherson'.

already: interpolated in MS.

week: ensuing word 'had' deleted in revisal of MS.

of his room: interpolated in MS.

conclusions: substituted in late revisal of MS for 'result'.

And then there came the incident of the dog: these words may be a later interpolation in MS: it is hard to say, as space existed for their inclusion whether written immediately after their predecessors or later. In any case the remnant of this MS page ('21') has been cut away. It is possible that the dog's death was not present in the Mordhouse version: [p.] '22' of the MS has evidently been deleted from what is now MS p. '26'. This MS page begins with Inspector Bardle's visit to Holmes at the point when he states Murdoch is the only suspect. This may mean that the deleted and destroyed half-page simply introduced the inspector's visit much as we have it now (a few strokes from the tops of letters of the first line in the torn-away portion are consistent with 'Inspector Bardle' in ACD's handwriting, although they are so few that we can put it no stronger).

183 *faithful*: interpolated in MS.

an Airedale terrier: a rough-coated dog, named after the region of Yorkshire where it was first bred. The breed of dog ACD had at the time, its inspiration is obvious. But the tradition of a dog's fidelity to ground fatal to its master, touching in the abstract, absurd in the concrete, was immortalized in the Edinburgh of ACD's boyhood when in 1871 a fountain was placed outside Greyfriars Kirkyard commemorating the terrier 'Bobby' who supposedly lived on his master's grave until he died: the landmark would have met ACD's eyes very often as closely adjoining the new Royal Infirmary and two of his own residences (1877–81). The present story may reflect medical student scepticism as to the dog's motives for remaining in the graveyard: it now seems doubtful if its master really was buried there, although the dog certainly lived there.

184 *and yet I could not*: changed in MS from 'and yet my mind would not be at rest'.

'Meaning Mr Ian Murdoch?': the original version, i.e. the Mordhouse version, now continues, with the likelihood that the paragraph and a half before this line were present in some form in the destroyed half-page '21'.

185 *in his mind*: A sentence constituting a paragraph was then deleted in the MS: 'It was clear to me however that a premature arrest would be fatal.' (In the Mordhouse version, evidently, Holmes has neither sought nor found a book and has no alternative suspects to Ian Murdoch other than (possibly) the Bellamy father and son, or—taking into account Murdoch's unexplained visit to the Bellamy house—a conspiracy of all three.)

bear in mind: substituted in MS for 'consider'.

Finally: substituted in MS for 'Then'.

with which these injuries were inflicted: substituted for '—the scourge—or whatever inflicted the weals'.

'*What could it be . . .*': seven previous lines deleted in revisal of MS:

> 'Have you heard the Doctor about that?' asked the Inspector with a peculiar look.
>
> 'Has he anything fresh[?]'
>
> 'He considers that it was not a mere wire whip which could have done the mischief. He has examined the scars very carefully with a lens.'
>
> 'What then?'

This passage would seem to have been deleted immediately on composition, or upon resumption of writing after a break.

185 *the doctor*: in the MS, not changed, 'the Doctor'. The only survival of that unborn Doyle hero Dr Mordhouse in his printed texts, it now must be taken to allude to an unnamed police pathologist.

186 *extravasated*: a fluid is said to be extravasated when it is spilt out of its proper vessel.

Have you?: inserted in MS revisal.

Perhaps I have. Perhaps I haven't: The Mordhouse version ran: 'Neither I am bound to say have I. And yet it is of capital importance...' Presumably on the deletion of Mordhouse, this became: 'Perhaps I have. Perhaps I have. My mind may be clearer soon...' That last sentence was then altered in MS to '... I may be able to say more soon...' No further MS changes were made at this point.

Anything... criminal.: in the MS 'clearly' is deleted before 'define', after which 'what made that mark' is substituted for 'the instrument used' and 'bring us a long way' for 'help us'.

comparison: Substituted in MS for:

> suggestion. But, as you say, the idea cannot be seriously entertained. What about a wire scourge with little bands of metal upon the wires. Have I not read somewhere of such a scourge.

cat-o'-nine-tails: a whip of nine knotted tails or lashes (once used for punishment in the army and navy and also an instrument of judicial sentence for grave crimes well into the present century).

Or... Mr Bardle.: Substituted in MS for: 'Yes, yes, but how does it advance us? The one argument for an arrest is that you could search the man's possessions.'

for an arrest: Interpolated in MS.

...'Mane'.: A further section of the original MS has been removed here, and this time it is far less easy to speak with confidence on its content.

...midday—?: It is probable that from 'Mane' to here was written after the deletion of Mordhouse, the paragraphs after here being the Mordhouse version as amended.

187 *Stackhurst*: substituted in MS for 'the naturalist Dr Mordhouse', whose hatlessness was noteworthy given his invariable sunhat.

For God's sake: interpolated in MS. Since the death of Sir George Newnes (1851–1910) it had become possible to use such language in the *Strand*.

at the sight: substituted in MS for 'in amazement'.

reticulated: in the form of a network.

death-mark: 'mark' interpolated in MS, probably very late.

It was half . . . pain: substituted for 'He was asleep.'

condition Stackhurst turned upon me . . . What is it?': substituted in MS for:

> safety I turned upon the Doctor.
> 'Well?'

In the first post-Mordhouse revision Stackhurst's 'What is it?' is simply repeated. His vocative 'Holmes' is a very late interpolation.

188 *over*: 'about' in book texts, possibly to lessen the assertion of nudity.

Indian file: single file.

I had: 'Then suddenly my' deleted as preceding words in MS.

a shout of triumph: From 'where did you find him' to here is evidently a newly written version taking the place of a passage now destroyed in which Dr Mordhouse took the lead to the lagoon.

'Cyanea!': some experts suggest that the jellyfish was not *Cyanea capillata* but the Portuguese Man-of-War.

I cried: 'he cried' in Mordhouse version, altered in MS.

The strange object at which I pointed: substituted in MS for 'It'.

rocky: interpolated in MS.

It pulsated with a slow heavy dilation and contraction: probably added during MS revision.

Its day . . . for ever: Substituted in MS for:

> 'Its day is done!' cried the naturalist.
> 'Help me, Holmes! Let us end it for ever.'

189 *flapping*: substituted for 'projecting' in MS.

thick oily: 'yellow' between them deleted in MS.

189 *It may have been the South-west gale that brought it up*: Interpolated in MS. This paragraph and its predecessor are a fragment

written evidently in place of an earlier Mordhouse version passage torn away.

When we reached my study: the next seven paragraphs are evidently the Mordhouse version, amended; 'study' is a substitution for 'house' in the MS.

pangs: substituted in MS for 'pains'.

Here is . . . ever dark: Substituted in MS for: 'Our explanations were unnecessary for Dr Mordhouse had arrived with a chocolate-backed book in his hand.' After the disposal of *Cyanea* he evidently went home to get it. Before 'when we reached my study' is a fragment ending a paragraph with the words 'became clear'—presumably spoken by Mordhouse and referring to the book.

Wood. Wood: separated in MS originally with 'said he' and close and open quotation marks respectively for first and second Wood. John George Wood (1827–89), a distinguished popularizer of natural history whose other work included *Common Objects of the Seashore* (1857), was the author of *Out of Doors: A Selection of Original Articles on Practical Natural History* (1874) and Richard Lancelyn Green points out that while the first edition was in green cloth it was reissued in chocolate cover binding (new editions 1882, 1890).

cobra: Next sentence deleted in MS.:

> I will leave the book that you may read the full account for yourself. [This remark is evidently made to Holmes, whom Mordhouse has so signally outgeneralled: nor can we be absolutely certain the Inspector is still present in the Mordhouse version at this stage.]

Could . . . described?: apparently an interpolation subsequent to the elimination of Mordhouse, and much more Holmesian in tone than the rest of the passage.

own encounter: 'fearful' between these words deleted in MS.

190 *disturbed*: substituted in MS for 'free'.

Inspector: substituted in MS for 'Mr Holmes'.

only: substituted in MS for 'nearly'. This sentence would seem to be the last known passage from the Mordhouse version, but the fair copy of many of the remaining paragraphs, in what at other points is most exceptionally detailed in its correction, indicates that much of what follows would have had earlier existence in that draft.

191 *concert-pitch*: the high-point of tension to which musical instruments are tuned for performance purposes.

Inspector, I often ventured: 'you have nearly met me at my Waterloo' deleted from MS between 'Inspector' and 'I', and 'you have nearly met me at what might have become my Waterloo' substituted only to be deleted. 'Often' beginning new sentence transposed in MS to be placed after 'I have'; 'have' retained in MS and in *Strand*, but subsequently deleted, perhaps in error.

Scotland Yard: Hitherto this last sentence would have seemed excessive: Holmes had been on the verge of much greater blunders in the past than that in the story as printed. But we may now see that the final sentence is a private joke, alluding to Holmes having actually had a very narrow brush with defeat at the hands of a rival, although one no longer in existence in the final version. It is almost a statement by the character that he had never been in more serious danger from his creator, even at the several attempts to end his existence by the writing of no more stories. It raises the suspicion that the story really was conceived to have Holmes record his own Waterloo. Whether the author privately repented of his intention, or was induced by his wife or by his editor to alter his decision, cannot now be said, and will probably never be known; but it would seem that the Mordhouse version had been brought to a level of considerable sophistication, if not fully refined in style, before its obliteration. We cannot state, for want of evidence, that Holmes's remark about his brain being like a crowded box-room is in fact his in the original draft, or whether a statement in such direct defiance of his view of the human mind as given in *A Study in Scarlet* was originally assigned to his victor, the naturalist, in order to present a figure sufficiently at odds with Holmes's own attitudes. It is clear that Holmes as a sportsman took his defeat well—his anxiety to tell the story against himself shows that—and it also explains *why* he should have wanted to single out such a case in his rare attempt at authorship. But it seems equally evident that the appearance of Mordhouse would have been a Waterloo indeed.

THE RETIRED COLOURMAN

First published in the *Strand Magazine*, 73 (Jan 1927), 3–12, with 5 illustrations by Frank Wiles. First American publication in *Liberty*, New York, 3 (18 Dec. 1926), 7–11, with 4 illustrations by Frederick Dorr Steele and map.

192 *Lewisham*: then a middle-class residential suburb of south-east London.

193 *within two years*: this dates the adventure in 1898.

Coptic Patriarchs: the Copts are people living in Egypt who practise their own form of Christianity. Their church is governed by a Patriarch or principal bishop.

It was late that evening: probably the first time that Watson fails to recount his own part in a case in direct statement to the reader (the self-revealing *Hound* documents not excepted)—a further distancing of the audience from its old narrator.

196 *Haymarket Theatre*: opened in 1720; the street called Haymarket was named after the market for hay and straw held there during the reign of Elizabeth I.

my old school number: in boarding schools each pupil is given a number, for purposes of identification, which he keeps throughout his school career.

197 *London Bridge*: at this time the main terminal of the London Brighton and South Coast Railway.

a Masonic tie-pin: incorporating an emblem of the Freemasons, a semi-secret fraternal organization with a large complement of professional and business men.

the gay Lothario: seducer, lady's man; named after a character in Nicholas Rowe's play *The Fair Penitent* (1703).

198 *Thanks to the telephone*: the telephone is also mentioned in two other *Case-Book* stories, 'The Three Garridebs' and 'The Illustrious Client'.

Carina sings to-night at the Albert Hall: not an operatic performance, given the venue, but a song recital. Hence this cannot be the comic opera *Carina*, as suggested by Redmond (*Sherlock Holmes: A Study in Sources*, 223), but must be his other candidate, the Venezuelan child pianist Maria Teresa Carreño (1853–1917) who became an unexpected success as a soprano from 1872 to 1889, when she returned to playing the piano and even did some composing and conducting. She sang the Queen in Meyerbeer's *Les Huguenots* (see last notes to *The Hound of the Baskervilles* in the present series) in Edinburgh on 12 Mar. 1872, which ACD must have seen or had reported to him. In fact she was presented as the advertised singer, Mme Colombo, who had walked out on the touring company performing at the Theatre Royal, Carreño being the solo pianist on the concert tour. The bogus prima donna was well, if not rapturously, received. The episode was naturally remembered in Edinburgh musical circles once it proved to have been Carreño's *debut.* (Marta Milinowski, *Teresa Carreño 'by the grace of God'* (1940)). The Albert Hall, a vast elliptical building of brick with terracotta decorations used for public entertainments, was begun in 1867.

betimes: early.

199 *Little Purlington*: like 'Mossmoor', a fictitious village.

Frinton: a coastal town in Essex. ACD stayed there in August 1913.

Crockford: *Crockford's Clerical Directory*, published regularly since 1858, is an alphabetical list of the clergy of the Church of England.

Living: benefice, parish from which a clergyman derives his income.

cum: (Latin) combined with; used when a country parish or living is made up of more than one village.

Liverpool Street: London terminus of the Great Eastern Railway.

200 *telegraph office*: Watson's instinct was to use this rather than the telephone.

201 *A white pellet*: probably potassium cyanide (to which, as a colourman, Amberley would have had access).

202 *and in order*: 'let all things be done decently and in order' (1 Corinthians 14: 40). It is to be hoped that Holmes does not mean the full paraphernalia of condemned cell and gallows.

203 *Broadmoor*: an asylum for the criminally insane in Berkshire.

204 *hermetically sealed*: closed off so that it is air-tight.

Burglary has always been an alternative profession . . . the front: a final salute to ACD's brother-in-law, Ernest William Hornung

(1866–1921), whose first collection of stories about Holmes's burglarious counterpart A. J. Raffles, *The Amateur Cracksman* (1899), was dedicated 'To ACD, This Form of Flattery' (with reference to 'Imitation is the sincerest form of Flattery', Charles Caleb Colton [?1780–1832], *Lacon*, I. 217).

204 *plastered rose*: perhaps more properly 'plaster rose'—decorated plasterwork shaped like a rose.

205 *foregathering*: in the sense of a chance meeting.

206 *on the body*: deleted from the *Liberty* text though retained in American book texts. The deduction is far-fetched, but not wholly implausible in that Josiah Amberley might be expected to get rid of his victims' personal effects with their bodies, and so could conveniently put the pencil in Ray Ernest's pocket. But it was far below his general level of intelligence not to work out why the pencil was on the floor near the body.

swank: ostentatious showing-off.

207 *North Surrey Observer*: a fictitious newspaper.

the true story may be told: the newspaper story and the self-serving MacKinnon are a reprise of the end of *A Study in Scarlet*, not so much suggesting an 1880s date (though see the note on Carina/Carreño, above) than acting as a celebration of the altruistic spirit of the Holmes cycle. This probably supplied ACD's reason for placing the story at the end of the book edition.

THE VEILED LODGER

First published in the *Strand Magazine*, 73 (Feb. 1927), 109–16, with 3 illustrations by Frank Wiles. First American publication in *Liberty*, New York, 3 (22 Jan. 1927), 7–10, with 4 illustrations by Frederick Dorr Steele.

208 *scandals*: examples of those Watson did publish may be found in 'The Noble Bachelor' and 'The Beryl Coronet' (*Adventures*), 'The Naval Treaty' (*Memoirs*), 'The Second Stain' (*Return*), 'The Bruce-Partington Plans' (*His Last Bow*), and 'The Mazarin Stone' in the *Case-Book*.

cormorant: a bird trained in the Far East to catch fish for its owner.

209 *forenoon*: Scotticism for 'morning'. ACD retained his Scots accent throughout his life but his reversion to early vocabulary in his final writing years has interesting parallels in other emigrants.

South Brixton: Brixton is a district in London, south of the Thames.

his tin: at this time milk was not delivered in bottles. The milkman carried on his cart a large churn, from which the milk was drawn in a small can or tin.

210 *Abbas Parva*: fictitious village in Berkshire, embodying a joke name that maliciously feminizes and belittles the Khedive Abbas II of Egypt (who gave trouble to the British during ACD's visit in 1896); the diminutive ('small female') contrasts with his formidable predecessor, Abbas Pasha.

211 *Buddha*: founder of the great Eastern religion that bears his name; often represented as sitting cross-legged, in an attitude of meditation.

the rival of Wombwell, and of Sanger: George Wombwell (1778–1850), owner of Wombwell's Travelling Menagerie, and 'Lord' George Sanger (1825–1911) and his brother John (1816–89) were noted nineteenth-century circus proprietors. George Sanger was shot by an employee at Finchley.

Sahara King: the Sahara desert occupies an extensive area of north Africa; the name is, perhaps, an association of ideas with Abbas Parva.

porcine: pig-like.

212 *Leonardo*: Leonardo da Vinci (1452–1519), Renaissance artist, polymath, and inventor, seems the origin of the name, given the Latin *vinco* (I conquer) and the inventive ingenuity.

Griggs: from the 18th-century proverb 'merry as a grig', for the irony of which see Mrs Ronder on p. 216. A grig is a grasshopper, or a sand-eel, or a cricket; thus a small lively person. The name seems an obvious derivation from 'Little Smil-ax' (see Appendix).

van: caravan.

Allahabad: city in north-eastern India, capital of the United Provinces of Agra and Oudh. The transfer of English provincial policemen to India must be accounted fairly unusual and would be more characteristic of the Royal Irish Constabulary.

213 *shutting the door*: the narrator actually does save his own life from a man-eating feline by this means in ACD's 'The Brazilian Cat' (*Round the Fire Stories*, 1908).

214 *the dear departed*: perhaps a reference to the one-act play of that (ironic) title by the Manchester dramatist Stanley Houghton (1881–1913), very popular at this time.

Montrachet: a wine from the Burgundy region of France (both t's are silent).

215 *madam*: Eugenia Ronder is British. Mrs Robert Ferguson and Isadora Klein are both 'madame' (being Peruvian and Spanish respectively). The distinction is not made in the earlier stories, where all married ladies are addressed as madam, including Mrs Beryl Baskerville, a Costa Rican (*Hound*); Professor 'Coram's' Anna, a Russian ('The Golden Pince-Nez', *Return*); and Signora Emilia Lucca, an Italian ('The Red Circle', *His Last Bow*).

217 *the Angel Gabriel*: in the Bible, Gabriel is mentioned as one of the seven archangels whose task it is to carry out God's immediate purposes. According to St Luke, he announced to the Virgin Mary that she would give birth to Jesus.

zinc: widely used at this time in the manufacture of such articles as pails and baths because of its resistance to corrosion.

219 *Margate*: seaside resort on the north Kent coast.

There was: the British (but not the American) book text inserts here 'on it'.

Prussic acid: hydrocyanic acid, a deadly poison; colourless with the distinctive smell of almonds.

SHOSCOMBE OLD PLACE

First published in the *Strand Magazine*, 73 (Apr. 1927), 317–27, with 5 illustrations by Frank Wiles. First American publication in *Liberty*, New York, 3 (5 Mar. 1927), 39, 41–2, 51, 52, with 7 illustrations by Frederick Dorr Steele.

The last Sherlock Holmes story had more baptismal trouble than most of its predecessors. It was announced at the end of 'The

Retired Colourman' (*Strand*, Jan. 1927) as 'The Adventure of the Black Spaniel', and while next month, after 'The Veiled Lodger', its title was given as 'The Adventure of Shoscombe Old Place' (two months before publication, though doubtless the copy was already in the *Strand*'s hands), the MS is reported (Baring-Gould, ii. 642) as having been headed 'The Adventure of the Shoscombe Abbey'. The conjecture that the *Strand* found this too close to 'The Adventure of the Abbey Grange' (*Return*)—presumably fearing that readers might suspect a reprint or cannibalization—and changed it is, borne out in part by ACD's using the MS title in his introduction to *Sherlock Holmes: The Complete Short Stories* (1928). He also, most singularly, there dated its appearance not one year earlier but three (1925), just as he dated 'A Scandal in Bohemia' (*Adventures*) a year later in the same passage. Possibly this was in subconscious defence against the 'decrepit gentlemen' cited in the preface to the *Case-Book* (q.v.).

220 *glue*: glue can be positively identified only by chemical means.

in the field: in the area under study through the microscope.

epithelial scales: flakes of dandruff.

St Pancras: probably the great Gothic Midland Railway Station opened in 1868, rather than the London borough.

the zinc and copper filings: expert opinion differs on whether these could have been identified by visual means.

the microscope: in none of the earlier Baker Street stories is a microscope mentioned.

half my wound pension: i.e. 'I have recently plunged too heavily in racing bets', not a confession of hopeless gambling as some commentators have suggested (as baseless as other anti- Watson charges of alcoholism, polygamy, etc.). The usage is a typical exaggeration in the style of the period. No mention of Watson's gambling appears at any other point: in 'Silver Blaze' (*Memoirs*) it is Holmes and not he who knows about the horse and who bets on a race whose outcome is irrelevant to the case.

'Handy Guide to the Turf': William Ruff (1801–56), sporting reporter, published an annual *Guide to the Turf* 1842–54, long imitated after his death.

Well, I should say so: Watson's up-to-date knowledge of Norberton arises from close consideration of his entry's chances for the Derby, on which Watson might now fear he has plunged too far. Hence his emphasis here. ACD is depicting the typical ceremonial punter, good for something on the Derby and the Grand National, and possibly something big. It is characteristic of Watson to plunge occasionally, and ponder subsequent disquieting information.

220 *my summer quarters*: presumably in his army days, although *A Study in Scarlet* seems to locate these entirely in India. A summer medical student *locum* may be intended (as in ACD's own experience). In any case, this would offer the casual punter a sentimental or nostalgic reason for choosing a Derby runner on which to bet, Shoscombe Prince reminding Watson of pleasant days at Shoscombe Park: this seems the only reason for introducing this first visit into the story, as it plays no subsequent part in the plot.

221 *Curzon Street*: named after George Augustus Curzon, Viscount Howe; a street in Westminster linking Park Lane to the west with Half Moon Street to the east. Clearly a snob money-lender.

Newmarket Heath: the name of a major racecourse, near the town of Newmarket in Suffolk.

buck: a fashionable young man, usually aggressive or extrovert.

the days of the Regency: from 1810 to 1820, when Britain was ruled by the future George IV, as Prince Regent, owing to the mental instability of George III. See ACD's pre-Regency novel *Rodney Stone* (1896).

plunger on the Turf: someone who gambles recklessly on horse-racing.

down Queer Street: in debt.

stud: stables where horses are bred.

222 *Harley Street*: runs parallel to Baker Street; was, and is still, a fashionable consulting-room address for doctors and surgeons.

the Derby: named after its founder in 1780, the twelfth Earl of Derby; the most important annual flat race for horses in the United Kingdom; run on the course at Epsom in Surrey.

up to the neck: i.e. in debt.

touts: people who try to obtain or spy out prior information about the horses running in a race.

223 *spins*: daily outings to exercise horses.

the Jews: the moneylenders to whom Sir Robert is in debt.

Crendall: fictitious place-name.

dropsy: the accumulation of watery fluid in body cavities.

227 *upper condyle*: the human femur (thigh-bone) has a condyle (protuberance) only on the lower (knee) end. Watson may have meant simply the upper end of the femur.

tend: *Strand*; 'attend' in British (but not American) book texts.

228 *'halt-on-demand' station*: a railway station at which trains stopped only if prior notice had been given to the guard.

Josiah Barnes: the last Josiah of the Sherlock Holmes cycle, which has surely the largest aggregate of persons of that name in literature—e.g. Josiah Brown in 'The Six Napoleons' (*Return*), Josiah H. Dunn in *The Valley of Fear*, and of course the retired colourman Josiah Amberley. The original King Josiah of Judah reigned 'for thirty and one years in Jerusalem' (2 Kings 22–3, 2 Chronicles 34–5), 'and he did that which was right in the sight of the Lord', in marked contrast to his predecessor and successor.

230 *out of the question*: Watson's disclaimer here cannot be snobbery, as has been suggested, since his case-notes abound with aristocratic miscreants. ACD would appear to be depicting a Derby plunger hoping against hope that Shoscombe Prince would not be scratched, as would happen if Norberton were arrested for his sister's murder and his creditors then foreclosed. This would account for Watson's earlier, and uncharacteristic, refusal to respond beyond a wooden 'I can make nothing of it' to Holmes's enticing 'vaguely sinister flavour'. He cannot find the thought of Norberton as a murderer inconceivable since he has already told Holmes that 'Norberton nearly came within your province once ... He nearly killed the man.' The mood of reprise in some final stories may be present here also, recalling Watson's awareness in *The Sign of the Four* that Holmes's success in investigations might be fatal to Watson's hopes.

a carrion crow ... eagles: the crow is a scavenger, living off the corpses of other creatures; the eagle—a bird of prey—is considered a nobler creature.

230 *dace*: small freshwater fish.

spoon-bait: a metal lure in the shape of a spoon which revolves when drawn through the water.

jack: young pike.

231 *heraldic griffins*: mythical monsters, each having a lion's body, feet, and claws combined with an eagle's head and wings.

barouche: four-wheeled carriage with a seat for the driver and two double seats inside.

Dogs don't make mistakes: dogs (except greyhounds) are said to have exceptionally poor sight. Holmes means that they do not make mistakes in smell (save for excusable errors such as that of Toby's in *The Sign of the Four*, Chapter 7, 'The Episode of the Barrel').

232 *groined*: in architecture, a groin is the curve or ridge formed by the intersection of arches crossing one another at right angles.

233 *Saxon*: North German race that conquered and settled in Britain during the fifth and sixth centuries AD, as described by ACD in 'The First Cargo' (1910).

Norman Hugos and Odos: names frequent among the Normans who invaded and settled in England during the eleventh century, under the leadership of William the Conqueror, whose half-brother Odo (d. 1097), Bishop of Bayeux, accompanied him in 1066 to be rewarded with Dover Castle and the earldom of Kent. Hugo of Avranches (d. 1101), possibly William's nephew, contributed sixty ships to the invasion and received the earldom of Chester; he conquered North Wales and Anglesey, becoming known as Hugo the Wolf. ACD presumably had both of these in mind, especially the somewhat Baskervillian Hugo.

jemmy: short crowbar.

234 *Baronet*: a hereditary knight. The first baronets were created after 1603 by James I.

sarcophagus: tomb, with pious legend and statuary work.

I could act no otherwise: '*Ich kann nicht anders*' (Martin Luther (1483–1546), speech at the Diet of Worms, 18 Apr. 1521).

235 *a dark horse*: i.e. one whose possibilities are known only to its owner and trainer.

my estate: as Watson and other interested parties already know, it is not Norberton's estate but that of his sister's brother-in-law, but the high hand of the bankrupt aristocrat is a vital element in his perpetual game of bluff. Compare Sir Lothian Hume in *Rodney Stone*.

237 *our humble abode*: the last recorded words of Sherlock Holmes are a final assertion of an old theme—unpretentious middle-class productivity against pretentious aristocratic unproductivity. The Saxon and Norman Shoscombe sees its dead dug up and burned in the interest of its class, while 221B Baker Street remains true to itself.

scatheless: unharmed, unpunished.

promises to end: 'ended' in the *Strand*. Presumably, the honour received by the old age is not in Shoscombe Old Place.

honoured old age: there are interesting parallels between 'Shoscombe Old Place' and its Irish-located predecessor of some forty-five years earlier, 'The Heiress of Glenmahowley' (*Uncollected Stories*), including an informative inn-keeper as to doings in the big house, whither the narrator and friend illicitly trespass in pursuit of adventure.

APPENDIX

A Source for 'The Veiled Lodger'

THE LOVE-LY TAM-ER, THE CRU-EL LI-ONS, AND THE CLEV-ER CLOWN

A TALE FOR THE LIT-TLE ONES

Written and Illustrated by
WALTER EMANUEL [1869–1915]

THERE was once a love-ly tam-er named Za-za, and some cru-el li-ons named Fi-do, Em-ily, Li-on, Kru-ger, Jane, Cæs-ar, and Rough, and a clev-er Clown named Lit-tle Smil-ax.

They were all in a Cir-cus.

One eve-ning Za-za was not feel-ing well. She had a bil-i-ous head-ache. She said to the Own-er of the Cir-cus, 'I feel ill. Need I go in-to the cage of Li-ons?' The Own-er, who was a cru-el man, said, 'You must. The Pub-lic must not be dis-ap-point-ed.'

So brave Za-za went in, but this eve-ning she had no pow-er ov-er the Li-ons. They re-fused to o-bey her, and sud-den-ly Cæs-ar and Rough rushed at Za-za, and knocked her down. 'Come on, oth-er li-ons,' said Cæs-ar, 'now we have her.'

Ever-y-one was in shrieks.

'Fetch red-hot po-kers,' cried the Own-er. But no red-hot po-kers could be found.

'Oh dear, oh dear,' cried Lit-tle Smil-ax, for he loved Za-za. Then a beau-ti-ful smile light-ed up his face. He had thought of some-thing. He ran swift-ly, and fetched his sham red-hot po-ker. Then he rushed to the cage with it, and when the cru-el li-ons saw him, they cried, 'Oh, look out, here comes a red-hot po-ker!' and they left Za-za, and ran to the far end of the cage.

So Za-za was saved, and a clerg-y-man was fetched, and Za-za mar-ried Lit-tle Smil-ax, and they had ev-er so man-y lit-tle clown-lets and col-um-bines, and the cru-el li-ons were pun-ished by hav-ing no pud-ding with their din-ner for a whole week.

(*Strand*, Sept. 1914)

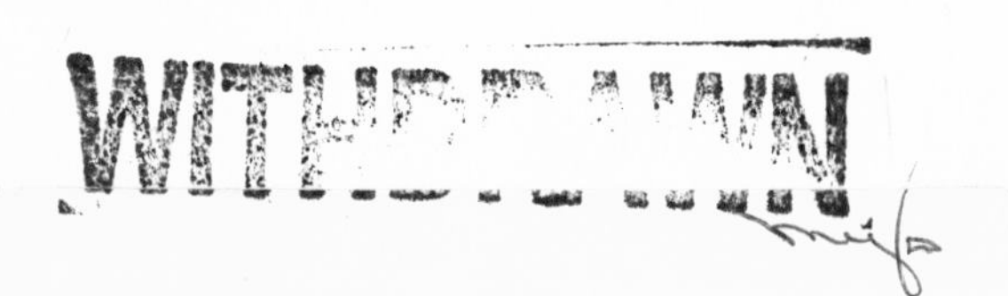